Mediating Sustainability

Kumarian Press

Selected Titles

Reasons for Hope: Instructive Experiences in Rural Development
Anirudh Krishna, Norman Uphoff, Milton J. Esman

Reasons for Success: Learning from Instructive
Experiences in Rural Development
Norman Uphoff, Milton J. Esman, Anirudh Krishna

The Human Farm: A Tale of Changing Lives and Changing Lands
Katie Smith

Achieving Broad-Based Sustainable Development: Governance,
Environment and Growth With Equity
James H. Weaver, Michael T. Rock, Kenneth Kusterer

Nongovernments: NGOs and the Political Development
of the Third World
Julie Fisher

When Corporations Rule the World
David C. Korten

Getting to the 21st Century: Voluntary Action and the
Global Agenda
David C. Korten

Beyond the Magic Bullet: NGO Performance and
Accountability in the Post-Cold War World
Michael Edwards, David Hulme

Democratizing Development: The Role of Voluntary
Organizations
John Clark

Intermediary NGOs: The Supporting Link in Grassroots
Development
Thomas F. Carroll

Mediating Sustainability

Growing Policy from the Grassroots

Jutta Blauert
Simon Zadek
editors

Kumarian Press

Mediating Sustainability: Growing Policy from the Grassroots.

Published 1998 in the United States of America by Kumarian Press, Inc.,
14 Oakwood Avenue, West Hartford, Connecticut 06119-2127 USA.
www.kpbooks.com
(860)233-5895   fax (860)233-6072

Production supervised by Jenna Dixon
Copyedited by Linda Lotz   Typeset by CompuDesign
Text design by Jenna Dixon   Proofread by Beth Richards
Index by Barbara DeGennaro
The text of this book is set in 10/13 Adobe Sabon.
The display type is ITC Officina Sans.

Printed in Canada on acid-free paper by
Transcontinental Printing and Graphics.
Text printed with vegetable oil-based ink.

∞ The paper used in this publication meets the minimum requirements of the
American National Standard for Information Sciences—Permanence of
Paper for Printed Library Materials, ANSI Z39.48-1984.

Library of Congress Cataloging-in-Publication Data
Mediating sustainability : growing policy from the grassroots / Jutta Blauert, Simon
  Zadek, editors.
        p.   cm.
  Includes bibliographical references and index.
    ISBN 1-56549-082-7 (cloth : alk. paper). — ISBN 1-56549-081-9 (pbk. : alk.
paper)
    1. Sustainable development—Latin America. 2. Sustainable agriculture—Latin
America. 3. Mediation—Latin America. I. Blauert, Jutta. II. Zadek, Simon.
  HC130.E5M423   1998
  338.1'88—dc21
  98-3325

07 06 05 04 03 02 01 00 99 98   10 9 8 7 6 5 4 3 2 1   1st Printing 1998

Contents

vi ■ Contents

Illustrations

Tables

Figures

Preface

Students in a master's program on environmental issues in Latin America were presented with a cartoon depicting two rural people standing on a mountain, question marks surrounding their heads, looking up to an image of suited men sitting around a table with voice bubbles stating, "sustainable development—blah, blah, blah." This was a simplistic way of inviting students to voice what they thought "sustainable development" meant, and who was defining it.

That was two years after Rio, and many of the students could not give a clearer idea about sustainability than the puzzled mountain people in the cartoon appeared to have. Today, the concept—and supposed practice—of sustainable development has been in the news for a number of years. Yet the interpretations of this slogan are still varied, often conflictive, and definitely confused. In some instances we are talking about largely irrelevant language differences. In most cases, however, this dense web of complex nuances reflects very real and different perceptions, values, and interests.

We have here, then, an issue of communication and interpretation, of priorities and values that differ according to each social group and context. Nonetheless, in the sphere of sustainable agriculture and rural development (SARD) and the era of Local Agenda 21 the term "sustainability" has found resonance with a range of people who have an interest in change in the countryside and the central issue of rural livelihoods. These people consciously seek out meeting "spaces"—be they concrete actions or planning and evaluative meetings. Opportunities are sought for clarifying independent positions and political spaces to win over others, tell others what to do, listen to other people's views, and even reach agreement over major or minor points for joint action. Yet negotiation between different points of view—the *mediation*—is still rarely leading to the fast and thorough structural change in SARD policy and practice that is required. Where these people are not finding an adequate mode of communication, failure leads to further violence or to unjust and environmentally damaging production and consumption patterns.

There are, however, changes afoot, albeit limited. New spaces are emerging, as the case studies in this book indicate, in Latin America. However, it is not yet easy to find a free flow of constructive exchange among people in the triangle of people-centered, conservationist, and economist approaches that in some sense delineates our three "bottom lines"—the social, the environmental, and the financial or economic. Different language needs persist and will not dissolve just as local identities and language rights are being fought over.

Is there a way out of this Tower of Babel? Can constructive policies really emerge if grassroots experiences and lessons learned by trading organizations, for instance, are not communicated? If farmers do not want to talk to those trying to establish fairer trade relations, and governments are content to rely on participatory rhetoric to win communities' votes rather than seeking to strengthen local capacity for development planning, then better practice at the micro or macro level will be difficult to achieve. Can a regional farmer organization aiming to develop a marketing strategy that aids sustainable livelihoods succeed if it does not understand the relevant meaning of "sustainable" to the local producer families? How else can such an organization understand which kind of market relations will benefit the community except from the community's perspective of what counts?

Our students were not alone, then, in being puzzled and insecure. What the ILAS (Institute of Latin American Studies, London University) and then a number of other institutions made possible was for us to organize a series of meetings between both old and new acquaintances to share experiences and doubts—both theoretical and practical—about paths toward SARD in Latin America. The Study Group on Mediating Sustainability was hosted by the ILAS and then the IIED (International Institute of Environment and Development) in London over eighteen months in 1994–95; several participants continued to collaborate in research, reflection, and action in 1997, and new collaborations have been set up. The six meetings were designed to be widespread. In London, meetings were held at ILAS, IIED, and the New Economics Foundation (NEF); in the Netherlands, a meeting was combined with a public conference at Wageningen University. The last meeting took place with some participants (both researchers and farmers) from Europe in Managua, Nicaragua, at the invitation of the National University of Agriculture and merged into a conference organized there.

People from different backgrounds (institutions, disciplines, political viewpoints, ages, and genders) were invited to contribute to areas of policy, practice, and methodological aspects that related to the theme of mediating, a concept that everyone appeared to endorse but about which there were few common actions or definitions. It was clear that the economists would speak differently from the activists, and the social scientists working with farmers would not initially understand the traders or the officials from multilateral agencies. In fact, many times a blockage in the desire to listen reflected the real-life constraints outside the meeting rooms. The willingness to rediscover some way of moving things forward, of untangling some of the threads, however, made this loose grouping of gloriously different people want to struggle with their insufficiencies and learn from one another. No new theories were developed; no solutions were offered or public declarations made that would change the world and miraculously make SARD happen for Latin America. Rather, doubts were raised, criticism handed out (mostly, but not always, constructive), and research results shared; papers were discussed, improved on, and commented on by those who could not attend, and collaborative work in the field emerged between a number of participants.

In practical terms, there was a set of core participants, the majority of whom had valid and relevant experience in issues relating to SARD in Latin America. Beyond this, a wider group of people from Latin America and elsewhere was involved, people who contributed particular insights at specific times to the process of the study group (for a listing of participants, see "Contributors and Study Group Participants").

The group first identified two issues surrounding SARD and communicating for change:

1. Among a small circle of researchers, nongovernmental organizations (NGOs), and planners favoring sustainable agriculture, the relevance of low-external-input agriculture and sustainable rural development may well be obvious. But how much have these projects and the sustainable agriculture movement actually influenced policy? Can experiences at the micro level be translated to policymakers at municipal, regional, and national levels? Where this has taken place, has the local project maintained the independence and specificity it desired?

2. At the same time, little attention has been paid to evaluation by the participants themselves—evaluation of the dynamics *within* their projects or organizations that have made their work successful, or of how they dealt with problems and mistakes. Self-evaluation often takes place internally in corridors and even in group meetings, but little is shared about internal decisionmaking structures and about the role and work of coordinators—the leadership qualities. Donors may request impact assessment studies, but little work can be found on the qualitative processes or on systematic and constructively critical analyses of impacts on the environmental, social, and economic relations of the community in which a project is located.

Having decided not to discuss these issues in the context of commercial agriculture, nor with regard to specific developments within and by the multilateral institutions (from the United Nations' Food and Agriculture Organization [FAO] to international research organizations such as the Consultative Group on International Agricultural Research [CGIAR]), the study group defined a number of objectives:

■ To identify obstacles to policy change for SARD at the macro and micro level.

■ To identify strategies for changing policy and attitudes that create such obstacles, with a particular focus on the need to understand communication issues and obstacles to change for different levels of decision making and action.

■ To analyze experiences in methodological directions for "bridging spheres." How are people influenced at the micro and macro levels, and how can transmission occur (scaling up)?

The format of the study group was academic, but also experiential and strategic. Participants offered papers broadly within an academic mold, but they drew directly on their experiences, whether at the field or the policy level. Each paper was intended to interrogate the central theme and to contrast and/or blend with other experiences and perspectives offered. We hope that the set of experiences and analyses brought together in this book reflects a range of specific examples of how practice is relevant to and can be turned into policies and practice that can underpin other people's initiatives and effectiveness.

The book is organized in four parts. First, however, the introduction presents the broader issues relating to sustainability and mediation, setting the context for the debate on Latin America. Central themes of communication, language shifts, market relations, and organizational implications are addressed to provide a guide to the diverse form and range of the chapters that follow. It summarizes the territory that is covered in later chapters and offers some personal views that are shared to varying degrees by most of the contributors to the book.

Subsequent chapters echo these themes but approach them from their own angles. Some conclusions relevant to mediation in terms of method, actors, and outcome are not always made explicit and do not adhere to a common definition. This diversity represents the different perspectives of each case study and analysis. The chapters also do not argue that one technology or method is the best (for example, whether participatory approaches for sustainable agriculture guarantee equitable land-use patterns); rather, they present examples of praxis and method that have given rise to mediation or methods for mediation.

In Part I, Pichón and Uquillas provide a wide-ranging overview of the contemporary history and current issues relating to small-scale agricultural production and sustainability within the context of technology development and policy responses from two of the most powerful actors in this sphere: the World Bank and the international and national agricultural research center sectors. Together with Bebbington's chapter on NGOs, Part I brings together two distinct spheres and actors of mediation: NGOs and governmental agencies and policymakers. This is not to argue that these two actors are the primary mediators. However, they are key actors in the practical and policymaking sphere. Being aware of the changes within these two realms allows a more acute understanding of the relevance of methodological and praxis-related action and mediation presented in the following parts. In addition, in these two cases, institutional responses are presented in the face of changing experience and political demands for new relations and approaches to achieve sustainable rural development. Active participation by these institutions and the farmers themselves in the practical and policy spheres, it is argued, is crucial for the success of mediation. Opting out by organizations and marginalization of the central constituency are no longer feasible, and this brings a new set of responsibilities.

The chapters in Part II address the practical issues of measuring—monitoring and evaluating—SARD initiatives. If mediation is desired, and if we want to find out where and how it has been effective, then methodologies are required that allow a better understanding and a communication of the different perspectives that stakeholders in SARD projects may have. The methodological examples all argue for inclusion of the quantitative with the qualitative, approaching systematization from different disciplinary angles.

Part III is concerned with the economics and organizational experience of small producer groups. Management of the markets and the international lobbying networks and new forms of trading relations have allowed mediation between some Latin American producers and the global trading system. Essential for this mediation is a recognition of the links with Northern trading and lobbying organizations, which can act as mediators themselves.

Organization building and the skills and personal transformations required for maintaining and using new alliances are the theme of Part IV. Whereas the Mexican experience echoes the importance of a commercial product in the struggle for SARD from Part III, the Central American farmers' movement opens a window into the personal development of leadership and mediation skills acquired by commercial and subsistence farmer organizations of the region. The social anthropology of new mediators for SARD provides the human voice of organizations.

The conclusion presents a synopsis by the editors of the various arguments made by the contributors. Summaries of the chapters are provided not to assess their accuracy or appropriateness but to assist the reader in seeing the key contributions made by each chapter to the central theme of the book. Beyond this, each chapter must be judged by the reader on its own merits, as must the book's success in fulfilling its aim of contributing to an understanding of this critical issue.

Together, it is hoped, the chapters make a coherent whole, though not in any sense do they to offer a picture of a seamless reality or of a foregone conclusion. The editors have played a cautious role in helping to construct a picture of events and conclusions together with the authors of these stories. We have built on the wealth of information and insight provided in the various papers. We have, above all, drawn on the extensive and often intensive discussions that took place over two years. We have benefited from the input of an enormous number of colleagues and friends. Finally, we have sought to make sense of often muddled conversations about strikingly obvious points, or attempted to penetrate our often worryingly simple views about what are clearly complex topics and issues.

To achieve all this through the process of this study group would be success indeed. Maybe it was a somewhat overambitious vision held by the instigators of and other participants in the study group, all of whom have committed time and considerable effort to this process. Of course, other people are also engaged in exploring this theme—practice to policy. We hope that this book contributes to their work as much as theirs has already contributed to ours.

Acknowledgments

The life of the study group that gave rise to this book has been one of flexibility, moves, and many utopian ideas not quite but almost made real. Thanks are due to all those who contributed to the process that formed the bedrock of this book.

As always, institutions were essential in creating space for activities: the Institute of Latin American Studies in London provided space and initial funding for the first two meetings, and the International Institute of Environment and Development in London offered similar support for further meetings and seemingly endless photocopying. The staff of the Institute of Development Studies (IDS) of the University of Sussex helped get the final manuscript out of a complicated computer pipeline.

The offices of the New Economics Foundation and Farmers' Link helped with the two last meetings in the United Kingdom. The conference and workshop in Wageningen were funded generously by the Dutch Social and Economic Science Research Council (CERES). We are also grateful to British Council offices in Mexico and London and to the British embassy in Bolivia, which covered travel costs of several study group participants. The other participants covered their costs from institutional and personal resources. To all of them, our sincere thanks.

This study group was a team effort on all fronts. Thanks are extended to Alberto Arce and Sietske Schoute, who organized the conference and workshop in Wageningen with great efficiency. Alberto also was mediator for the participation of study group members at the conference in Managua in July 1995, and Carlos Barahona and Alistair Smith handled the brunt of the last months of coordination from the U.K. side. For the conference of the PASONAC (Central American Program of Sustainable Agriculture), Mauricio Rodríguez and Alberto Sediles, the vice-rector of the Nicaraguan National Agricultural University, performed the roles of hosts and coordinators, as well as fund-raisers. The papers of that conference were published by the university in late 1995 as *La Universidad Nacional Agraria Coadyuvando al Desarrollo de una Agricultura Sostenible*, a tribute to the efficiency of the resource-starved Central American university staff. Some of those papers we hope to include in a Latin American version of this book; Smith's paper is already a chapter in this book.

Not all participants contributed chapters to this book, which was neither intended nor possible. This book could not, however, have come into being without the contributions of study group members who commented on others' papers and enriched them with their discussions. Several colleagues contributed their original material in written format to others, but their contributions are not directly represented in this book. These people shall not remain invisible, particularly those who gave constant input into other people's papers—not a common characteristic among academics or NGO activists. The editors and study group coordinators, above all, could not have done without them: Simon Anderson, Roberto Escalante, Alberto Arce, Marcel v.d. Does, Steve Wiggins, Denise Humphreys, José Antonio Péres, Graham Woodgate, Kees Blokland, Peter Gubbels, Abraham van Eldijk, Paul Engel, Ruerd Ruben, and Araceli Burguete. Isabel Carballal gave days of free labor for the assiduous transcription of tapes of the two last meetings in the United Kingdom—a generous piece of work—as was the contribution by participants who typed up their minutes to be shared among all.

Again, the fun of networking was enhanced by the material provided by selfless and interested external participants and supporters. Lack of funds or lack of time was the principal reason for them being unable to attend meetings directly. Their contributions—through comments on papers and research proposals—were of great importance to the debates, to the professional development of participants, and to the exposure of many Latin American organizational processes that we otherwise might have stayed ignorant of. They are Rubén Pasos, Frans Doorman, David Kaimowitz, Rusty Davenport, Osvaldo Feinstein, Carlos Brenes, Gonzalo Tapia, Raúl Hopkins, Rosario León, and Alistair Smith.

Also, our thanks go to James Lupton who applied his cool-headed skills to much of the text and his generosity of spirit to the editors, both at the last minute; and to Linda Beyus, Briana Rosen, and Jenna Dixon at Kumarian Press for their patience and diligence as our publishers and editors.

It was not easy for all involved—as consultants, trainers, academics, students, and activists—to find the time to come together. We all benefited from this particular mix of perspectives and styles. It is hoped that this book will be felt by study group participants—distant or close—to be *their* book. To us, it feels like a window into reflections that public conferences would not have permitted, and into critiques and self-doubts that we all are still trying to manage in our daily work in the field, in the classroom, at the negotiating table, or while sweating over empty sheets of paper.

About the Editors, Contributors, and Study Group Participants

Editors

Jutta Blauert, Fellow at the Institute of Development Studies at the University of Sussex, England, is a rural sociologist, working as a trainer and researcher on participatory evaluation and monitoring methods in the context of sustainable agriculture and integrated rural development programs. In Mexico, she is attached to the Research and Postgraduate Studies Center in Social Anthropology (CIESAS) in Oaxaca.

Simon Zadek is research director of the New Economics Foundation, London. As an economist and political scientist, he has worked for the last fifteen years in policy advice and organizational development.

Contributors

Gerardo Alatorre is an agronomist with postgraduate training in rural development and social anthropology, working with the Mexican NGO GEA-Pasos.

Anthony Bebbington is a human geographer, lecturing at Colorado University, Boulder, USA.

Eckart Boege is a social anthropologist at the Veracruz region's National Institute of Anthropology and History and is director of a sustainable community development project in the Calakmul Biosphere Reserve, Mexico.

Marc Edelman is a social anthropologist and associate professor at Hunter College and at the Ph.D. Program in Anthropology, both of the City University of New York.

Irene Guijt, a tropical land and water use engineer, works as a research associate in the sustainable agriculture program of the International Institute of Environment and Development, London.

Francisco J. Pichón is a social scientist at the World Bank's Environmental Unit of the Latin America Office.

Marion Ritchey-Vance was originally trained in philosophy and linguistic and international relations. Between 1965 and 1996 she worked with the Inter-American Foundation in many Latin American countries and as coordinator for learning and evaluation.

Alistair Smith is a development educator and *animateur*, with training as a linguist and in development studies. He works for the Norfolk Education & Action for Development Center in the United Kingdom and its associate organizations Farmers' Link and the Banana Action Network.

Pauline E. Tiffen is a political scientist and a managing director responsible for trade development activities of TWIN, a London-based charity working toward fair trade. She is also a trader, as a member of the managerial staff of TWIN's associated company, Twin Trading.

Jorge E. Uquillas is a sociologist and rural development specialist, working at the World Bank's Environmental Unit for the Latin American Technical Department.

Other Study Group Participants* (1994–96)

Core Participants

Simon Anderson (Department of Agriculture, Wye College, London University)
Alberto Arce (Department of Sociology, Wageningen University, Netherlands)
Carlos Barahona (Department of Statistics and Center of Agricultural Strategy, Reading University, UK)
Isabel Carballal (master's student, Wye College, London University)
Alasdair Cook (veterinary faculty, Merida University, Mexico and Wye College, London University)
Marcel vander Does (Department of Sociology, Wageningen University, Netherlands)
Roberto Escalante (Department of Economics, National University, Mexico)
Denise Humphreys (Peru Office Coordinator, Inter-American Foundation, Arlington, Virginia)
Bernadette Keane (master's student, Wye College, London University)
David Preston (Centre for Development Studies, University of Leeds, UK)
Birte Rodenberg (Institute of Development Sociology, Bielefeld, Germany)
Colin Sage (Environment Section, Wye College, University of London)
Steve Wiggins (Department of Agricultural Economics and Management, Reading University, UK)
Graham Woodgate (Environment Section, Wye College, University of London)

External Participants

Carlos Brenes (FAO, Forest, Trees and People, Central American Coordination Office, Costa Rica)

---

*Core participants attended all or most meetings and presented papers. External participants attended one meeting or gave substantial input (comments, papers, material) from a distance. Occasional participants attended one meeting or presented a paper.

Russell Davenport (freelance consultant in rural development and evaluation, San Francisco)

Frans Doorman (ETC Foundation, Leusden, Netherlands)

Osvaldo Feinstein (Monitoring and Evaluations Division, International
Fund for Agriculture and Development [IFAD], Rome)

Raúl Hopkins (Economics Department, Queen Mary and Westfield
College, London University)

David Kaimowitz (IICA, San José, Costa Rica)

Rosario León (FAO, Forest, Trees and People, South American
Coordination Office, Bolivia)

Deborah MacLauchlan (WWF Latin America Desk, UK)

Rubén Pasos (FUNDESCA/CADESCA, Managua, Nicaragua, and Panama)

José Antonio Péres (Centro de Estudios y Proyectos, La Paz, Bolivia)

Michel Pimbert (WWF International, Biological Diversity Division and
Agriculture Working Group, Geneva)

Ruerd Ruben (Programs of Peasant Economics and Sustainable Land Use
and Food Security, Wageningen University, Netherlands)

Occasional Participants

Kees Blokland (Paulo Freire Foundation, Arnhem, Netherlands)

Aracelí Burguete (Independent Federation of Indian Peoples, Mexico)

Abraham van Eldijk (Department of Agricultural Law, Wageningen University, Netherlands)

Paul Engel (Department of Communication and Innovation Studies,
Wageningen University, Netherlands)

Peter Gubbels (regional coordinator West Africa, World Neighbors,
Burkina Faso)

Sabine Gündel (Agriculture and Extension Department, Free University,
Berlin, University of Yucatán, Mexico)

Renwick Rose (WINFA, St. Vincent, Caribbean)

Others attended the public conference at Wageningen University in
November 1994 and the International Seminar on SARD at the National
University of Agriculture in Managua, Nicaragua, in June 1995, especially
the Central American speakers:

Antonio Belli (SIMAS)

Roberto Blandino (FACA-UNA)

Marta Gutiérrez (CARE-UNA)

Allan Hruska (CARE-UNA)

Orlando Núñez (CIPRES)

Javier Pasquier (CIPRES)

Abelardo Rivas (UNAG)

Carlos Zelaya (FARENA-UNA)

Zaida Zúniga (CARE-UNA)

Introduction

The Art of Mediation:
Growing Policy from the Grassroots

Jutta Blauert and Simon Zadek

> We are facing new forms of engagement, where unlikely alliances bring
> unexpected returns, where old coalitions prove to be rigid and often
> counter-productive, and where the traditional opposition often seems to
> be saying just what you are saying whilst meaning something altogether
> different. (Zadek and Blauert 1995, 7)

The future of both agriculture and rural communities in Latin America will
depend on whether existing and emerging policies support or undermine
practices consistent with principles of sustainability. For this reason, rural
communities seek to influence the policies of the state, of commercial organizations, and of international agencies. To this end, rural people and farmers'
organizations form alliances with a range of actors, notably nongovernmental organizations (NGOs), but also, increasingly, research institutes, international donors, and the business community.

Alliance building in itself is not new. However, today's alliances are quite
distinct from alliances for survival in previous times. History has shown
farmers, particularly subsistence and indigenous farmers in Latin America,
that alliances with partners from centralized political parties or even military
regimes have rarely improved the long-run situation of resource-poor and
politically weak rural families. The impact of structural adjustment policies
in Latin America, for example, may have led to some improvement in
macroeconomic indicators, but poverty has been exacerbated in the process.
In 1994, poverty in Latin American was higher than in 1980, and rural
unemployment has increased alongside the decollectivization and privatization of productive lands (Scott 1996).

Today, rural actors are increasingly aiming to strengthen their roles in
critical policy formation processes. This assertive agenda is addressed by
forming alliances that allow them to develop and apply alternative measurement systems and research methodologies, to form and operationalize new
consultative mechanisms, to demand effective adjustments to credit policies,

and to develop policy-oriented institutions controlled by farmers, rural communities, and their supporters.

This book explores ways in which rural communities in Latin America have sought to influence policies affecting their livelihoods and the quality of their natural environment. A selection of experiences from different countries highlights the importance of collaboration between rural communities, producer organizations, NGOs, and advisers. The experiences offered map out a potential range of common themes and strategies, providing insights into the means by which *practice* can influence *policy*. Although the diversity and complexity of policy formation processes make it difficult to generalize, it is possible to identify patterns that offer lessons that can be used to strengthen initiatives in the future.[1]

The following chapters describe initiatives that seek to distill and articulate knowledge emanating from the *realm of practice* of sustainable agriculture and rural development (SARD) in a manner that can influence the *realm of policy*. The objective is not to analyze the specific changes in policies of governments, donors, or NGOs.[2] Rather, we focus on how particular forms of language used to describe practice—and the individuals and organizations that mold and project those language forms—have evolved to support the process of influencing policy.

This process of seeking points of action, of forming alliances, of influencing, and of facilitating and analyzing praxis toward the goal of SARD is the sphere of what we call *mediation*. This process requires paths or channels. It is these languages, people, and organizations that make up the channels through which this influence can be brought to bear. These, then, are what we mean by the *mediators*. Finally, the use of language, the communications media, and new institutional arrangements are the tools that are deployed by organizations and people as active *agents* of change.

This book is, above all, about mediation between different worldviews and interests. However, whereas language was previously one means by which to distinguish these different actors and perspectives, this is now rarely the case. For example, farmers' organizations and multilateral lending organizations often remain, as in the past, antagonistic toward one another. Yet today they organize their arguments around similar terminology, symbolized best by the common use of the term "sustainable development" and its ever-widening range of derivatives.[3] Furthermore, these actors are now increasingly talking to one another whereas political animosity and traditional patterns of racism and other forms of bigotry had in the past limited or prohibited any direct contact. It is not argued, therefore, that there is *one* best or politically correct mediator, organizational type, or mediation tool.

The newly emerging patterns of communication that the contributions to this book refer to do not imply that the old problems have disappeared.

"Networking" for marketing and lobbying and using state-of-the-art communications technology still must be complemented by the practice of steady, daily listening and learning about the perceptions of other stakeholders, be they scientists, policymakers, or subsistence farmers. Following the argument made by Does and Arce (1995) that we can advance more effectively toward the objective of SARD if we learn to be more sophisticated in our reading of project narratives, our argument here is that *mediation* involves a careful listening to stories represented by others—stories that have historically been ignored or misheard within the development industry.

A case can be made for cautious optimism. Public-sector agencies and even corporate bodies are increasingly acknowledging, willingly or under duress, the legitimacy of the interests and rights of farmers, rural communities, and related actors in agricultural and rural development processes. Yet this acknowledgment does not do away with vested interests and the power struggles underlying policies formed on the back of consultative processes (Anderson 1996). Simply communicating more or using the right "sustainability" or "participation" language is not sufficient. Mediation processes between different conceptions of sustainability should not be confused with consensus, the amelioration of conflicting interests, or the alleviation of poverty. The emergence of new or renewed arenas for "interinstitutional concertation"[4] is occurring at a time when real conflicts of interest are becoming accentuated as a result of macroeconomic policies associated with trade liberalization, privatization, and the retreat of the state. Moderate—and moderated—mediation is not a substitute for the more familiar forms of confrontational campaigning on social justice, human rights, and environmental issues, which certainly remain necessary. The experiences described in this book have occurred within this context, in recognition of and in response to the deterioration of the quality of life and opportunities of many rural communities throughout Latin America (Oxfam 1996; see also Chapter 1).

Images of Globalization

Agricultural development in Latin America is generally perceived as a problem area in terms of sustainability. Rates of degradation, excessive dependence on oil-based inputs, and uncertain or costly credit, coupled with uncompetitive pricing in the international markets, have cost the majority of the region's population dearly over the last fifteen years. Rural poverty is on the increase across the region, as is rural out-migration. The social and environmental pitfalls of the green revolution have been widely analyzed and documented (see Agudelo and Kaimowitz 1994; Linck 1993; Calva 1993; Chapter 1 of this book). Today, a dominant image is of the transnational company increasingly monopolizing profitable rural markets, from basic grains to piggeries and forestry. Rural communities and their supporters consider the current processes governed by liberalization as ones in which the continent is being "eaten up" by market forces that are uncontrollable by all but the strongest in the game of survival from the land.

Yet the traditional alternative to this scenario, the autarchic vision of organic agriculture and small-scale rural production systems, has so far been unable to deliver sustainable livelihoods for the majority or even large numbers of rural dwellers. Off-farm incomes are still (and often increasingly) more attractive in comparison to the low liquidity that much of sustainable small-scale agricultural production implies where markets are weak (see Ruben and Heerink 1996).

"Scaling Out"[5]

For practice to impact policy and to have an effective impact on individual and communitywide livelihoods, vertical interaction is not enough. Linkages need to be established horizontally; skills in the management of resources and markets need to be shared at the regional level—for instance, between farmers' organizations. "Scaling up" of impact and activities, as the experiences of social enterprises in the region demonstrate, needs to be coupled with regional alliance seeking and exchanges between producers. In addition, scaling up needs to go hand in hand with "scaling out," with a greater diversity in forms and a greater depth in quality of participation and engagement between different actors. This book takes as a starting point the view that isolated initiatives at the community level cannot prevent the process of globalization from undermining both social and ecological systems. For grassroots initiatives to offer more than temporary relief, they need to influence the broader policies that set the terms on which SARD will, in the longer term, stand or fall.

Limitations on the impact of initiatives operating solely within communities are reported from several cases within the participatory technology development sphere (see Chapter 4). To enhance effectiveness, it is argued that horizontal and vertical linkages between the community and other actors need to be established and maintained. Local initiatives are, indeed, increasingly making use of the actors at the meso and macro levels in the sustainable management of local resources, so as to strengthen regionalized and even transnational attempts to achieve social and ecological sustainability. These new approaches to communication and access to data are, therefore, an integral part of a strategic response to the same problematic circumstances. They represent serious attempts to influence the policy agenda and counteract negative globalization effects in favor of a sustainable development path (see Engel 1995; Engel and Salomon 1994). Globalization, furthermore, has taken place in ways and in arenas (media, marketing, social justice networks) over which rural organizations can gain some leverage and advantage. Rural organizations can and increasingly do, for example, actively use such trends to their advantage in counteracting international pressures that negatively affect the local SARD agenda.

Changing Institutions

Over the last decade, shifts in research and teaching institutions have also raised the prospect of mainstream national and international institutions accepting, or even proactively developing, public policy relevant to the needs of SARD.[6] SARD has traditionally been associated with two approaches: agricultural production and environmental stewardship, focused on low-external-input agricultural practices; and social production of knowledge and decisions, particularly focused on the principle and practice of participation. These two pillars of SARD are intimately related. Indeed, one of the most important successes of SARD initiatives has been to confirm and legitimize the knowledge of actors such as small-scale farmers in the practice of environmentally appropriate forms of farming. The now quite widespread recognition of the relevance of low-external-input agriculture practices has served to reconfirm that knowledge (see Scoones and Thompson 1994).

In spite of only limited experience in policymaking conducive to SARD, extensive experience by the NGO, community-based, and research sectors has left seeds for hope and resistance across the wider region (Carroll 1992; Kaimowitz 1995). Increasingly, the same research and development institutions that were in charge of applying and enhancing the green revolution, such as the Consultative Groups on International Agricultural Research (CGIARs), have moved toward the small farm sector and away from crop-specific research. The maize CGIAR, CIMMYT, in Mexico is a good example, having conducted research on nitrogen-fixing cover crops employed by several of the farmer-to-farmer agricultural and soil conservation programs in the region. The tropical tuber CGIAR, CIAT, in Colombia has long been undertaking research work with farmers, including studies of participatory methods encouraged by current rural development thinking, such as Rapid Rural Appraisal or Participatory Rural Appraisal (RRA or PRA). Supported by new policy drives within the Food and Agriculture Organization (FAO), the International Fund for Agricultural Development (IFAD), and the CGIAR training center the International Service for National Agricultural Research (ISNAR), these institutions have considerable influence over national agricultural resource centers and teaching institutions. In the end, it may well be through these institutions that individual policymakers learn of new practices and opportunities (FAO 1992, 1994a, 1994b; Scoones and Thompson 1994).

A number of factors, however, have tended to undermine the institutional efforts and resources required to secure the long-term stewardship necessary for effective SARD. Of critical importance is the continuing globalization of markets, particularly for agricultural products; the associated structural adjustment processes; and the reduced or changed roles of the state accompanied by a rising importance of NGOs. Technology requiring little external or energy-intensive inputs like many of those rooted in indigenous knowledge systems is an attractive solution in principle but will not withstand the pressures of globalization. Research focused solely on raising the technical efficiency of small-scale rural production processes is unlikely to be meaningful as a long-term guard to autonomous rural livelihood strategies. Anderson (1996), writing on the experience of a Mexican agricultural university, argues that the need for institutional support to create learning environments for SARD has been recognized by most research and teaching institutions, but it is still restricted in practice by institutional barriers, among them the reluctance of research councils to fund collegiate participatory research in the agricultural and rural development arena:

> So far, few agricultural research centres have made the policy changes necessary to provide the institutional framework for participatory research. . . . Professional rewards and career structures are not geared to encouraging participatory research and are not tolerant of the requirements for resources and time that such work implies. . . . A challenge faced by [practitioners of participatory action research and those using participatory rural appraisals] is to demonstrate the efficacy of their methodology to policy makers so that further adoption of the methodology might take place. The problem of the time scale of participatory projects from initiation to impact might make this task arduous. Policy makers often work to political deadlines that tend to be short term. Those National Agricultural Research Centres [that are] already operating successful participatory programmes and are aware of the restrictions imposed by economic policies, might facilitate the process of convergence between the interests of government institutions and participant rural communities. Here the advocacy role of participatory researchers is important. (Anderson 1996, 10, 13, 21)

The potential role of researchers as mediators through practical engagement and policy advisory functions has become stronger, as failures of the green revolution and technology transfer approaches have become more widely acknowledged.

Institutional opportunities and blockages are also encountered by actors within the public sector. Escalante (1994) reports that in the case of a Mexican governmental agency funding agricultural technology transfer and production projects, training in participatory learning and planning methods in the 1990s did not guarantee a changed practice on the ground. Training in participatory approaches to working with farmers by itself is no mediating tool, nor does it guarantee a move toward SARD supported by the institution in question. He cites some institutional and political constraints that echo those facing researchers: institutional and personal investment into participatory practice may not be given by the public sector, because "individuals in charge of relevant departments are motivated to achieve results which will justify their status[,] demonstrate that they follow government's macroeconomic guidelines[, and] keep their bargaining power in negotiating their institution's future budgets" (Escalante 1994, 6).

Sustainability Derailed?

The language of sustainability evolved in large part from the margin—albeit from politicized intellectuals rather than from rural communities themselves (Zadek and Blauert 1995). Sustainability—for the nonconservationist actors—reflects an agenda of social justice within and across current and future generations as well as environmental stewardship, whether as a means, an end, or some combination. In conceptualizing sustainability, there has been a focus on the interrelationship between its social and environmental foundation blocks of rural development strategies, resulting in an overall socioecological approach to understanding development processes. For the current debate in Latin America, sustainability has therefore come to embody an agenda that extends beyond economic viability and environmental regeneration, reaching deeply into the structure of social organization itself by insisting on the key component of social equity and justice. For many Latin American people, sustainable rural development often represents more a struggle for democratic rights and land rights than an ecological regeneration or conservation. However, Leff argues that

> [t]he environmental issue demands the preservation of the natural base of resources in the interest of a sustainable production. This implies the need to revalue the *ecological conditions of production* and to create the political conditions for a reappropriation of the [peasant farmers'] natural means of production. . . . This has been leading to new political strategies for the appropriation and socialization of nature and has generated new productive practices for a sustainable agriculture. (Leff 1996, 38–39)

Sustainability, in this sense, is also on the brink of being moved off this particular track. Often rhetoric is taken over by interest groups effectively repressing the economic organization of small producers, or else short-term economic and political interests overshadow the ecological interests of future generations. In these cases, new mechanisms for monitoring policy changes, new institutions, and relationships between these need to be found. A transparency in the evaluation and monitoring processes in the field and strengthened skills in resource mobilization are required to give different actors the legitimacy to make sustainability a social reality.

Over two decades, the language of sustainability has penetrated the policy discourse and become common currency. At exactly the same time, patterns of production and trade have threatened its attainment. Interpretations of what constituted paths to sustainable development have multiplied to include precisely the same paradigms that had originally been challenged and to exclude the voices and perspectives that had sought to present these challenges. Transnational corporations and dictatorships alike became the eloquent advocates of sustainable economic growth, while the needs of rural communities and environmental systems continued to be ignored and threatened (Korten 1995).

The mainstreaming of the sustainability debate has therefore been a mixed blessing. It *has* opened up new opportunities for influencing those institutions at the heart of the policymaking process. At the same time, it has created an often impenetrable cocktail of linguistic devices and consultative roller coasters that have absorbed and often weakened the capacity of progressive institutions struggling for meaningful progress: the growing of policy from the grassroots.

The challenge, then, is to find new mechanisms, or to strengthen those that already exist, in order to institutionalize at the policy level the advances made in *practice* by rural communities and their supporters, collaborators, and partners. This is the route from practice to policy. It is, to say the least, a difficult one to travel. However, the practices referred to in this book *do* provide an insight into an ever-growing countercurrent to the dominant approach to economic development. They are single experiences or case studies, but they have to be set alongside changes in policymaking, organizations, and technical expertise that have been described elsewhere.[7] These practices—organizational, productive, research, and methodological—have emerged from within weak social groups in internally stratified and often conflictive rural communities. Such communities have traditionally been informed of policy rather than consulted about it. The practices concern small-scale production processes, often with low levels of external inputs to suit the needs and capacities of resource-poor rural people. Where they relate to the business of markets and distribution, they often highlight the potential of the local or the national. When they concern international markets, they challenge in all cases the rights of large companies to dominate trade and to retain large portions of the overall financial resources generated.

These agendas and perspectives do not enter easily into conventional policy processes. Institutions with the power to influence policy rarely have a form or internal structure that allows for dialogue with, let alone real learning from, those below. This is certainly the case for most public bodies and commercial organizations, and it is also the case for many NGOs. The very structure of work and reward has historically constrained such a listening process and has prevented what is heard from being taken into account. The professions (academic, finance, or political), for example, generally do not reward their members in financial terms and status for communicating with rural people who have few academic qualifications, and they often speak the dominant language poorly or incompletely (Chambers 1993; Said 1994). These rural people are usually inhibited in communicating their extensive experience in agricultural techniques and processes in a manner that resonates with professional, institutionalized thinking, and they are therefore ignored as a matter of course, even when they are the central subjects of that discourse (Zadek 1993).

New Actors for Praxis and Mediation

A new set of actors emerged during the late 1980s and early 1990s in Latin America, partly as a response to this pernicious and systematic linguistic confusion. These actors were intent on asserting the core propositions of sustainability in policy forums. These actors saw their task as being able to "translate" the imperatives of SARD into a language that resonated and was therefore more effective in influencing the policy debate. Three key sets of institutional actors have taken on this role: NGOs and research and development institutions; public-sector and international funding agencies; and grassroots organizations, whether as farmer organizations, community-based organizations (CBOs), or community enterprises. In some instances, one point on this institutional triangle sought to influence others—for example, CBOs seeking to influence NGOs. In other cases, coalitions formed among the different types of institutions, such as NGOs and donors working to influence government agencies (for these common-cause coalitions, see Biggs and Smith 1995).

However, it was recognized that little could be achieved without shifting the terms on which poorer groups of rural producers worked within the market. A fourth set of actors has therefore entered the fray: the business community itself. This constituency has long been shunned by the development community as being essentially part of the problem, despite the fact that the business community has a pervasive impact on every community and in all likelihood will need to be co-opted into the sustainability agenda if it is to be effectively addressed—exactly because so much rural poverty and injustice used to be associated with the commercial sector. A growing and often unhappy acceptance of this fact has meant that there is an explosion of new relationships between private-sector and small-scale rural producer organizations. The aim, however, is to ensure democratic relations guaranteed by associated mechanisms and respect for the needs of rural people.

Most policies in the rural sector, designed to change farmers into small or medium entrepreneurs to feed an urban market, were criticized—often for good reason—for destroying the very foundations of the rural livelihood systems that had ensured peasant community survival for centuries. Long-term farmers' survival strategies used to be based on customs of exchange labor and consumption arrangements, low reliance on external inputs, and regenerative agricultural practices. These were effectively destroyed in most cases with agricultural modernization, damaging the production and marketing systems that had assured survival and identity at least on a minimal level (Kaimowitz 1995). Whether this somewhat idealized view of rural communities and their external enemies is correct or not, the dependence of rural people on trade and external relations is undeniable. Modernization in agriculture and high-external-input technology may have increased production, but they have not prevented a rise in rural poverty and rural-urban migration.

Successful and cost-effective implementation of a participatory approach requires the mobilization of the skills, talents, and labor of the rural population. To make this possible, decentralized administrative, fiscal, and political systems conducive to the genuine participation of rural people in decision-making, execution, and accountability processes are required. Involvement of the private sector is needed in this effort, but partial or full governmental financing is also required for a small-scale farmer-oriented strategy to achieve its natural resource management and poverty reduction goals. Public investment in traditional agricultural areas may have high opportunity costs relative to the areas of commercial agriculture where most agricultural growth has occurred. But efficiency trade-offs may be justified in terms of the greater impact on poverty and environmental degradation that can be achieved in traditional areas.

It is along these tracks, many contributors to this book argue, that a derailing of sustainability can be avoided.

Needing Mediation

To mediate is to arbitrate, moderate, facilitate, or even umpire or referee a process of dialogue between parties. Mediation in this sense involves a process of "coming between" different interests with a view to finding a way forward from what is, or is in danger of becoming, a cul de sac of conflict or inertia. Beyond this, however, lies a more partisan approach to mediation, which involves a decision to step in, intercede, and help to negotiate a process with an orientation toward particular interests. In its most transparent form, this type of mediation is effectively one of advocacy. In either case—or at any point on the spectrum between these two extremes—a mediation process can work only through the identification of some common ground on which a deal can be struck. Such common ground is more easily arrived at when the different sides have shared values and discover a language that expresses those values in a way that both can comprehend. Often, however, the identification of common ground requires more than this kind of fortuitous discovery. It requires pressure by one or both sides to shift the values of others, or at least to bring them into line with very different values and interests.

Sustainability has multiple interpretations, each constructed to support a particular agenda and legitimized through the use of particular ideological and methodological constructs. Mediation in this context is a facilitation of dialogue and persuasion not only between actors with conflicting aims but also between actors who defend themselves by recourse to the cause of sustainable development.

By *mediating sustainability*, therefore, we mean the way in which we and others can play a role in channeling the knowledge required to support the informing and influencing of those people involved in policy formation. These people certainly talk of sustainability, but they reflect very different life experiences and interests from those of people working and living at the grassroots level. The challenge is to be more effective in channeling those grassroots experiences and facts in a manner that enables policymakers to understand, believe, and act in a sane manner.

This book as a whole suggests that grassroots or community organizations are by no means the only ones that understand what is most likely to constitute a sustainable development path. However, the objective is acknowledged to be the contribution to a process whereby rural people are able to define and manage their own livelihood system while practicing a socially just and environmentally sustainable resource use. To achieve this aim, local worldviews, perspectives, and interests need to be made visible or brought forward so that they, along with insights gained at a more macro level, can be woven into relevant policy.

A key theme throughout this book is communication. Those adopting the role of mediator have sought to improve their effectiveness by developing and applying methodologies that aim to establish acceptable forms of communication—through both language and activities—to provide a bridge between people and institutions from micro and macro spheres. Mediation therefore requires the construction of a common basis—such as indicators agreed on by both farmers and policy institutions—that can be used to assess, for example, the effects of a controversial agricultural subsidy program such as PROCAMPO in Mexico (Gómez et al. 1993)[8] or infrastructure development by a social forestry enterprise in Costa Rica. In many instances, this involved the development of evaluation systems that could effectively transmit to policymakers the experiences of rural communities in seeking to implement SARD under such policies and to expose rural institutions to organizational and technical experiences and advances in different regions. In other cases presented here, this implied a demonstration in and through the market that certain forms of production and trading processes were viable as well as consistent with the principles of SARD. In all cases, mediation required the development of institutions through which critical messages could be channeled, refined, and strengthened. Institutional development today constitutes an integral part of a mediation process that aims to contribute to SARD.

In a sense, every institution and every person or group can potentially be a mediator in this sphere. The contributions to this book do not argue that one organization or person is *the* mediator for SARD. It is not up to just the NGOs or the experienced lobbyists to take on these roles. Of importance is the *approach*, the way in which and the direction toward which mediation takes place. The new environment for testing and designing rural and agricultural development alternatives also points to the need for skilled people who are trusted by all or at least several parties and can mediate constructively between opposing views. Effective methods or appropriate language adopted by mediators not only sends intelligible signals between the parties but also acts to highlight both the relevance of their own skills and their overall legitimacy. When researchers make use of a standard economic cost-benefit analysis, for example, they send certain signals that validate emerging perspectives to some while undermining their relevance or accuracy to others. When a nonprofit organization successfully markets a fairly traded product, the effective transmission of its message to commercial traders can be much greater than a more traditional verbal assertion of the need for justice in trade.

There are many elements required for the *effective* translation of practical experience into policy initiatives—for instance, clear conceptual models, an understanding of and ability to manage the necessary switches in language, and an understanding of and ability to cope with professional and institutional paths of policy formation. In addition, an astute analysis of political deadlocks is required, or of the violence hidden behind the language of the negotiation table (as happens in so many cases, from Mexico to Brazil to the lands of the Mapuche), in order to judge when and when not to move forward or to cede points. At every stage there are deals, compromises, and decisions to be made. All of these, and many more, are part of the process of mediating sustainability.

Toward Optimism

For this cautiously optimistic tale to be realized in practice requires new methods, tools, and techniques. Whereas some of these concern the practicalities of agricultural production processes and rural livelihoods, others concern the more ephemeral needs underlying the new communication imperatives that go hand in hand with engagement in negotiations with those who were previously enemies or simply unreachable or unknown (Winter 1995). Methods are also needed that enhance learning processes within each of the sectors involved (for example, peasant farmers, commercial farmers, policymakers)—that is, learning related to engagement with new contexts and new neighbors. The examples of regional interinstitutional development committees such as those in Bolivia (Péres 1996), the Chilean municipal planning committees, and the Mexican technical forestry committees support the relevance of bringing different stakeholder groups together in analyzing constraints and opportunities for new policies and practices.

Effective mediation toward the new millennium requires several key activities, as the lessons from Latin America and the Caribbean tell us (see Chapter 9).

Policies and Actors in Mediation

The spread in Latin America of participatory methods of planning for and evaluating SARD objectives is now being employed in the interest of mediation for locally defined rural development paths, as well as to strengthen local or regional economic initiatives. The Andean planning method used by NGOs such as COMUNIDEC, community-based land use planning in southern Mexico, and regional development plans in Bolivia or Argentina are just some examples. The new roles taken on by different institutional actors in this sphere, however, allow new lessons to be drawn. NGOs, public-sector agencies, and research institutions participate to mediate and use the participatory discourse to secure their own involvement in the rural sphere.

Measurement as Mediation

Mediation requires a "conveyor belt" of concise and relevant information. The use of indicators is a highly contested arena, and one in which Latin American institutions are contributing much to the macro-level agenda (see MacGillivray and Zadek 1996). But attempts at communication are still not effective enough in designing and *using* indicators with and between different stakeholders: NGOs or grassroots organizations still largely conceive measurement as a threat of external evaluation. In fact, the "indicator game" is too frequently used as a controlling evaluation instrument by donors and governments alike. Yet measuring is a powerful tool, and locally defined indicators are demanded to complement or replace some of the national-level indicators holding no meaning for local actors (see Blauert and Quintanar 1997).

Within the SARD arena, indicators have served two purposes. First, indicators allow change to be monitored, along with the relationship between these changes and defined goals or missed objectives (for example, a community-defined idea of sustainability for the village and immediate surroundings). Second, and closely related to the first, is that indicators can be a means for refining and articulating different perspectives on local as well as global sustainability, thereby offering a critical basis for different perspectives to be compared and related and providing a clearer sense of the position and performance of each party to negotiations. Indicators are, therefore, potentially both a means of evaluation and a means of mediating between different interests and perspectives; that is, they provide a tool for learning and for communication.

Markets as Mediation

SARD is not feasible without economically viable alternatives to approaches that are not socially and environmentally sustainable or attractive. Finding new ways of engaging in the market is increasingly recognized by many rural organizations as a key element of any strategy to strengthen their position not only economically but also politically. Mediation within the new era of international trading blocks—and within the context of initiatives to carve out more equitable and environmentally sound approaches to trade—demands steep learning curves by producers of cash crops and of manufactured goods alike. Niche marketing can have an impact on credit and export policies for smallholders that had not previously been thought possible. Local and regional seed banks, as well as genetic resource conservation, may come to be the alternative axes for mediation for subsistence producers in the region.

Organizing for Mediation

Mediation for SARD may increasingly be a skill of individuals who are able to bridge cultural, professional, and language gaps. Yet the rise of farmer organizations in Latin America in the commercial and policy arenas—however limited their power may appear—points to the need for regional approaches to mediation. Without a regional organization, coffee producers in southern Mexico would not have been able to win a position in the international market (see Chapter 6), nor would Brazilian rubber tappers have been sufficiently strong to negotiate their extractive reserves. The organizational skills of the peasant farmer producer sector are drawing on lessons from the commercial sector. Lobbying NGOs, in turn, are learning quickly to incorporate financial, managerial, and negotiating skills in their work, areas that had previously been the prerogative of the private commercial sector.

New strategies, then, require new capacities and capabilities. There is a need for conceptual frameworks, methods, tools, and tactics that fit the need to engage, communicate effectively, and ultimately persuade a range of people and institutions that are unfamiliar with, skeptical of, or downright opposed to the concerns and requirements underlying SARD (Biggs and Smith 1995). Some of these new modes of communication will be oriented toward the traditional sources of policy formation—institutions of the state—particularly those involved in agricultural research and development. However, there is an increased recognition of the need to find means of influencing economic processes through direct interactions with the corporate sector or through active involvement in markets. In all these cases, there is both the need and the opportunity for institutions of civil society to be innovative in their approaches to self-organization and alliance building so that they effectively support these new forms of engagement.

Notes

The authors express their gratitude to colleagues who commented on earlier drafts of this introduction and the conclusion: the participants of the study group, especially Irene Guijt and Pauline Tiffen, as well as Stephen Biggs, Michael Edwards, John Farrington, Jude Fernandes, Robert Potter, and the anonymous reviewers. The final responsibility for this text lies solely with its authors.

1. The study group did not discuss sustainable agriculture and rural development (SARD) in the context of large-scale commercial agriculture in the region. This book does not do so either. The reason for this lies as much in the professional background and interests of the contributors and study group participants as in the fact that most SARD initiatives (in practice and policy spheres) are concerned with small and medium-sized producers and the more fragile rural environments. The study group recognized the importance of the large-scale commercial agricultural sector for the long-term success of any SARD strategy that aims to improve environmental aspects of agricultural production, but it was felt that the alternatives to the failures of green revolution–based agricultural production systems and rural development models have come from the small farming sector. For this reason, and for the eminent validity of analyzing a production sector that is central to the vast majority of the population in the region (whether as urban consumers, migrants, rural dwellers, or farmers), this book concentrates on the small farming sector and on the dynamics of organizations of small commercial and subsistence farmers and their farming systems.
2. There are many publications that analyze of the impact of policies on rural people and the agricultural and environmental sectors (for further references, see Chapter 1; for Mexico, consult Grammont and Tejera Gaona 1996).
3. The term "sustainability" was until recently voiced predominantly by those wishing to challenge the dominant development model that takes economic

growth as the basic indicator of successful progress. "Sustainable development" in this form was the antithesis of a development model that considered that social and environmental problems would be solved by increasing the size of the economic "pie." Today, "sustainability" is a more explicitly contested term. It is still used to describe a vision of local economies using minimal resources to sustain small communities at modest levels of income and consumption. In addition, however, the concept is part of a formulation of "sustainable growth," where it is at best intended as a moderating adjective and more cynically is used to offset potential criticism of the continued dash for economic expansion.

4. Concertation, from the Spanish *concertación,* is a commonly used expression for political negotiation and consensus seeking between different actors. Originally, this word was used only by public-sector institutions, aided by multilateral funding agencies' language, in seeking the involvement of dissident rural and urban social actors in the proposed policy process. Today, even some NGO sectors accept this word as an expression of the endeavor to find a common purpose and dialogue between different actors in the interest of rural development.

5. Thanks to John Gaventa (1998) for the term, contrasting horizontal "spread" with "scaling up."

6. SARD is used as a framework for this discussion, since it describes the systemic view of natural resource management and agricultural production and marketing systems: agriculture is set within the sphere of broader rural development, the latter including trading and processing parts of a wider livelihood and policy sphere affecting the lives of all rural dwellers, not just the actual agricultural producers.

7. Practical experiences in the realm of social enterprise, sustainable agriculture, and rural agricultural knowledge systems, as well as methodological development, can be found in academic and practitioners' literature, such as the *ILEIA Newsletter* and *Forests, Trees and People Newsletter* (published by a program within the FAO). For Latin America, key references for analytical material of technical relevance that has managed to inform much of SARD policymaking include Bebbington et al. 1993 and Scoones and Thompson 1994.

8. PROCAMPO (Program of Support to the Countryside) is a public-sector program that pays small-scale farmers in Mexico a fixed annual sum per hectare of land planted with maize. It is a program of direct producer subsidy and encourages the production of the staple maize in resource-poor rural areas.

References

Agudelo, A., and D. Kaimowitz. 1994. Las Implicaciones Institucionales y Metodológicas de Promover un Patrón Tecnológico más Sostenible para la Agricultura. Una Reflexión con Base en Dos Casos de Colombia. Manuscript. San José, Costa Rica.

Anderson, Simon. 1996. Research Centres and Participatory Research: Issues and Implications (Mexico). Revised paper presented to the Study Group on Mediating Sustainability, Institute of Latin American Studies, London.

Bebbington, A., G. Thiele, P. Davies, M. Prager, and H. Riveros, eds. 1993. *Nongovernmental Organizations and the State in Latin America.* London: Routledge.

Biggs, Stephen, and Grant Smith. 1995. Contending Coalitions in Agricultural Research and Development: Challenges for Planning and Management. Paper presented to the conference Evaluation for a New Century: A Global Perspective,

Canadian Evaluation Association and American Evaluation Association, Vancouver, Canada, 1–5 November.

Blauert, Jutta, and E. Quintanar. 1997. Seeking Local Indicators—Participatory Self-Evaluation. *Appropriate Technology* 24 (2): 21–23.

Calva, José Luis. 1993. Principios Fundamentales de un Modelo de Desarrollo Agropecuario Adecuado para México. In *Alternativas para el Campo Mexicano*, edited by José Luis Calva. Vol. 2. Mexico City: Fundación Friedrich Ebert/UNAM.

Carroll, Thomas F. 1992. *Intermediary NGOs: The Supporting Link in Grassroots Development.* West Hartford, Conn.: Kumarian Press.

Chambers, Robert. 1993. *Challenging the Professions: Frontiers for Rural Development.* London: IT Publications.

Does, Marcel vander, and Alberto Arce. 1995. The Use of Narrative in Project Evaluation: A Case from Ecuador. Revised paper presented to the Study Group on Mediating Sustainability, Wageningen, Netherlands, November.

Engel, Paul. 1995. Facilitation Innovation: An Action-Oriented Approach and Participatory Methodology to Improve Innovative Social Practice in Agriculture. Ph.D. diss., Agriculture and Environmental Sciences, Wageningen University, Netherlands.

Engel, Paul, and Monique Salomon. 1994. RAAKS, a Participatory Action-Research Approach to Facilitating Social Learning for Sustainable Development. Chapter for the International Symposium on Systems-Oriented Research in Agriculture and Rural Development, Montpellier, 21–25 November. Also presented to the Study Group on Mediating Sustainability, Wageningen University, Netherlands, 17 November.

Escalante, Roberto. 1994. Participation and Economics: Illustrations from Mexico's Agricultural Sector. Paper presented to the Study Group on Mediating Sustainability, Institute of Latin American Studies, London.

FAO. 1992. *Políticas Agrícolas y Políticas Macroeconómicas en América Latina.* Social and Economic Development Study no. 108. Rome: FAO.

FAO. 1994a. *Participación Campesina para un Agricultura Sostenible en Países de América Latina.* Series Participación Popular, no. 7. Rome: FAO.

FAO. 1994b. *Sustainable Agriculture and Rural Development. Part 1: Latin America and Asia.* DEEP Series. Rome: FAO.

Gaventa, John. 1998. The Scaling-up and Institutionalization of PRA: Lessons and Challenges. In *Who Changes? Institutionalizing Participation in Development*, edited by J. Blackburn and J. Holland. London: IT Publications.

Gómez, C., R. Manuel Angel, R. Schwentesius, M. Muñoz Rodríguez, et al. 1993. *¿PROCAMPO ó ANTICAMPO?* CIESTAM, Universidad Autónoma de Chapingo, Reporte de Investigación no. 20. Mexico City.

Grammont, H. C. de, and H. Tejera Gaona, eds. 1996. *La Sociedad Rural Mexicana Frente al Nuevo Milenio.* 4 vols. Mexico City: Plaza y Valdés.

Kaimowitz, David. 1995. El Avance de la Agricultura Sostenible en América Latina. Manuscript. Instituto Interamericano de Cooperación para la Agricultura, San José, Costa Rica.

Korten, David C. 1995. *When Corporations Rule the World.* West Hartford, Conn.: Kumarian Press.

Leff, Enrique. 1996. Ambiente y Democracia: Los Nuevos Actores del Ambientalismo en el Medio Rural Mexicano. In *Los Nuevos Actores Sociales y Procesos Políticos en el Campo*, edited by H. C. de Grammont and H. Tejera Gaona. Mexico City: Plaza y Valdés.

Linck, Thierry. ed. 1993. *Agriculturas y Campesinados de América Latina. Mutaciones y Recomposiciones.* Fondo de Cultura Económica/ORSTOM, Mexico.

MacGillivray, Alex, and Simon Zadek. 1996. Medir la Sostenibilidad: Revisión Sobre el Arte de Hacer que Funcionen los Indicadores. *Investigación Económica 56* (218): 139–176.

Oxfam. 1996. *Poverty Report.* Oxford: Oxfam.

Péres, José Antonio. 1996. Reforms, Actors and Popular Participation in Contemporary Bolivia. Paper presented to the Study Group on Mediating Sustainability, Wageningen. Revised version.

Ruben, Ruerd, and Niko Heerink. 1996. Economic Approaches for the Evaluation of Low External Input Agriculture. Draft paper presented to the Study Group on Mediating Sustainability, London. (For a shorter version, see Economic Evaluation of LEISA Farming, *ILEIA Newsletter* 11 [2]: 18–20.)

Said, Edward. 1994. *Representations of the Intellectual: The 1993 Reith Lectures.* London: Vintage.

Scoones, Ian, and J. Thompson, eds. 1994. *Beyond Farmers First: Rural People's Knowledge, Agricultural Research and Extension Practice.* London: IT Publications.

Scott, Chris. 1996. El Nuevo Modelo Económico en América Latina y la Pobreza Rural. In *La Nueva Relación Campo-Ciudad y la Pobreza Rural*, edited by A. P. de Teresa and C. Cortes R. Mexico City: Plaza y Valdés.

Winter, Anne. 1995. Is Anyone Listening? Communicating Development in Donor Countries. UN Non-Governmental Liaison Service, Development Dossiers UNCTAD/NGLS/57, Geneva.

Zadek, Simon. 1993. Bridging Spheres of Communication: Information Exchange for Sustainable Land-Use. Report. Overseas Development Institute, London.

Zadek, Simon, and Jutta Blauert. 1995. Mediating Sustainability: Practice to Policy. Paper presented to the Study Group on Mediating Sustainability, New Economics Foundation, January.

Part I

Policies and Actors in Mediation

1

Sustainable Agriculture Through Farmer Participation

Agricultural Research and Technology Development in Latin America's Risk-Prone Areas

Francisco J. Pichón and Jorge E. Uquillas

Poverty-driven environmental degradation is common throughout Latin America and the Caribbean's heterogeneous natural resource base, with the most critical areas in northeastern Brazil, southern Mexico, and the densely settled hillside areas of the Andes, Central America, and the Caribbean. These areas contain the largest concentrations of rural poor in the Latin American and Caribbean (LAC) region.

In the poorest areas of Latin America and elsewhere in the developing world, the nexus of rural poverty, rapid population growth, and unsustainable agriculture is leading to the degradation of land, water, and forest resources, as well as to the breakdown of indigenous institutions and their natural resource management systems, which are critical for sustaining the livelihoods of the poor. Resource degradation in these lands also originates in the unequal distribution of land and other natural resources and in the economic and political marginalization of the people whose livelihoods depend on these resources. Policymakers cannot afford to underestimate the poverty reduction impact that agricultural development has in these degraded areas, especially among small-scale farmers. Since farmers and other resource users are the custodians of natural resources, agricultural development has a direct impact on how these resources are managed.

Agricultural intensification that increases the productivity of scarce resources is crucial to poverty reduction and improved natural resource management in these areas (Kevin and Schreiber 1992). Experience gained from the green revolution in parts of Asia has shown that broad-based agricultural growth involving small and medium-sized farms and driven by productivity-enhancing technology can offer a way to create productive employment and alleviate poverty on the scale required (Pinstrup-Andersen and Pandya-Lorch 1995). However, when inputs are mismanaged, new problems are created, such as increased pest resistance and a narrowing genetic base as large numbers of traditional crop varieties are replaced by relatively few modern varieties. When yields are raised by diverting more water for irrigation or by increasing the use of chemicals, there are also increased possibilities for waterlogging and salinization of irrigated lands, reduction of organic life, and chemical contamination of soil and water, with severe consequences for human health.

Various questions have grown out of the realization that resource-poor farmers stand to gain very little from the processes of development and technology transfer characteristic of the green revolution. Critics of the green revolution have pointed out that the new technologies were not scale-neutral. The farmers with the larger and better-endowed lands gained the most, whereas farmers with less resources often lost, and income disparities were often accentuated (Chambers and Pretty 1994). Although subsequent studies have shown the spread of high-yielding varieties and benefits for smaller farmers in green revolution areas where they had access to irrigation and subsidized agrochemicals, some disparities remain. Perhaps even more significant is that the areas characterized by traditional agriculture remain poorly served by the transfer-of-technology approach, due to its bias in favor of modern scientific knowledge and its neglect of local participation and traditional knowledge.

This chapter argues that if agricultural intensification is to be desirable and sustainable in poverty-driven, environmentally degraded areas in Latin America, it must be based on a more participatory approach for technology development and dissemination. More attention to agriculture and natural resource management in traditional risk-prone areas will help solve the problems of poverty, food insecurity, and environmental degradation. We argue that improvement of the management of natural resources is not only linked to the alleviation of poverty but also essential for achieving sustained productivity increases in traditional and ecologically vulnerable areas. In these regions, agricultural intensification may be the only realistic strategy for addressing poverty and environmental problems, including on-site erosion and desertification or off-site forest destruction due to expulsion of the population to the humid tropical lowlands. Such a strategy should include delineating an agenda for policy formulation for a participatory natural resource management practice based on both farmer-based traditional innovations and selective external inputs. To design and implement such an agenda, cooperation among governments, international agencies, nongovernmental organizations (NGOs), the private sector, and the technical and scientific communities will be required.

The first section of this chapter describes the problems of poverty, food insecurity, and natural resource degradation in the region's risk-prone areas and highlights the significance of agriculture for the development of these areas and the economies of the region. The next section develops a characterization of the agricultural intensification process in traditional risk-prone areas and in the higher-potential and commercially oriented agricultural areas. We discuss the possible linkages between responses to poverty, population pressures, agricultural intensification, and natural resource management in these areas, and their likely implications for the agricultural sustainability of the region. The third section discusses the achievements and limitations of modern green revolution technologies and the adverse sociopolitical and ecological impacts resulting from mismanagement of external inputs. The next section reviews some examples of traditional methods of natural resource management and discusses the opportunities in combining local farmers' knowledge and skills with those of external agents to develop and/or adapt appropriate farming techniques. The chapter concludes with a discussion of the main issues and building blocks to be considered in the development of a new strategy toward traditional risk-prone areas.

Poverty, Agriculture, and Natural Resource Degradation in Latin America and the Caribbean

The Problem of Persisting Rural Poverty

Poverty alleviation is one of the principal institutional goals of international development agencies such as the World Bank. However, during the past twenty-five years, the LAC region has made little progress in reducing the overall level of poverty. Over 46 percent of the population is still poor, and the number of poor has increased from 120 million in 1970 to 196 million in 1995. The rural population in Latin America is projected to remain stable at 125 million until the year 2000, whereas urban population will grow from 275 to 400 million. Although these trends should enhance the prospects of achieving more sustainable agriculture, there is still the problem of persistent rural poverty. Over 61 percent of the rural population is poor, compared with 39 percent of the urban population, but with increasing levels of urbanization, poverty is said to be assuming an urban face (CEPAL 1994).

Remarkably, advocates of the urban development strategy have ignored the fact that many of the urban poor grew up in rural areas and continue to maintain links to these areas. Much of the massive population transfer to the cities has been into the informal sector and slum areas, where migrants contribute to the urban poverty problem. Many others have turned to environmentally destructive cultivation of marginal lands and forest frontiers. As a result, poverty is increasingly becoming entrenched in neglected areas of traditional agriculture, and it is also acquiring gender-specificity as more poor households are headed by women, often because men migrate to forest margins or urban centers in search of employment.

Still, because 75 percent of the population in the LAC region now lives in urban areas and urban poverty is becoming widespread, it is assumed that poverty in the region is mainly an urban phenomenon. The aggregate picture is misleading, however. Four countries—Brazil, Mexico, Argentina, and Colombia—are large and relatively urbanized, and thus dominate the regional statistics. Brazil and Mexico alone account for 51 percent of the total area of Latin America, 54 percent of the population, and 58 percent of the region's agricultural gross domestic product (GDP) (World Bank 1993). The rural population accounts for 50 percent or more of the total population in Bolivia, Paraguay, Haiti, Guatemala, Honduras, El Salvador, and Costa Rica. In addition, a much higher proportion of the rural population is poor, even in some of the more urbanized countries. For example, Colombia's rural population is less than 42 percent of the total, but 74 percent of the poor live in these areas. Venezuela has a rural population of 16 percent, but 30 percent of the poor live in these areas (Wiens 1994).

Rural Poverty in Environmentally Threatened and Degraded Areas

Country-level empirical information on the dimensions and characteristics of poverty in environmentally threatened or degraded areas is scarce, although it is recognized that poverty-induced environmental degradation is a significant and persistent problem throughout the region. Wolf (1986) estimates that throughout the world, some 1.4 billion people, or about one-quarter of the world's population, lack sufficient income or access to credit to purchase appropriate tools, materials, and technologies to practice environmentally sustainable agriculture, protect natural resources against degradation, or rehabilitate degraded resources. These people have lost the capacity to support themselves in a sustainable way and live in what Chambers, Pacey, and Thrupp (1989) call complex, diverse, and risk-prone areas. Leonard (1989) refers to those rural areas as places of low agricultural potential and high ecological vulnerability, where limited soil fertility, adverse climatic conditions, or other natural factors inhibit the success of modern agriculture.

Although risk-prone areas are most widespread in Asia and sub-Saharan Africa, they are also found in northeastern Brazil and in many parts of Central America, Mexico, and the Andean region. It is estimated that 1 billion people live in risk-prone environments in Asia, 300 million in sub-Saharan Africa, and 100 million in Latin America and the Caribbean (Wolf 1986). At the same time, two-thirds of the world's 2 billion hectares of degraded lands are located in Asia and Africa, but poverty-driven degradation as a proportion of total agricultural land is most severe in the Andean region of South America and in Central America and Mexico, where one-quarter of the vegetated land is degraded (Oldeman, van Engelen, and Pulles 1990).

Because of deforestation, overgrazing, and other unsustainable agricultural activities, nearly 200 million hectares of land in Latin America are now moderately or severely degraded. It is estimated that severe soil degradation has affected 75 percent of agricultural lands in Central America alone. It is also known that most of the degradation in the LAC region is taking place in the breadbasket areas, densely populated rain-fed farming areas, and other areas providing important environmental services. Of the approximately 17 million hectares of forest that are cut down worldwide each year, 8 million hectares are in Latin America, with two-thirds of the land clearance in the region being for conversion to agricultural use by farmers. Latin America had the largest area of forest converted during the 1980s, reaching nearly 7.4 million hectares every year (Garrett 1995; FAO 1991).

The Importance of Agriculture in the Region

Of utmost importance to poverty and its relationship to agriculture and natural resource management is the question of food security and nutrition. Almost 60 million people in the region suffer from food insecurity, meaning that they do not get enough food to lead a healthy, active life. Six million of these are children. The number of undernourished people in the region was 55 million in 1985 and is projected to reach 62 million by 2000 (FAO 1992). Still, the image exists of a lessened role for agriculture in the economies of the LAC countries, especially as the region has become more urbanized. However, although agriculture contributes only 10 percent of GDP for the region as a whole, it accounts for over 25 percent of national GDP in Bolivia, Ecuador, Paraguay, Haiti, and Nicaragua.

In addition to the variable contribution of agriculture to GDP, the food and agricultural system, which includes agroindustry, accounts for 25 percent of all economic activity in the LAC region. The vitality of the food and agricultural system is even more important to the economies of the region's poorest countries, such as Bolivia, Guatemala, Honduras, and Paraguay, where there is pervasive poverty in the rural sector. It is also important to the more urbanized economies. Urban dwellers depend on agriculture for food and textiles, and a healthy agricultural sector generates employment in other sectors as rural incomes rise, creating demand for additional goods and services. It has been estimated that in Latin America every increase in agricultural output of US$1 increases overall economic output by almost US$4 (Garrett 1995).

The agricultural sector is also a more important source of export growth in Latin America than in other regions, accounting for 33 percent of merchandise exports. This share is nearly on a par with (but more diversified than) that of Africa, but it is significantly higher than that of East Asia (19 percent) and South Asia (24 percent). Another characteristic of agriculture in Latin America is the significance of livestock, perennial crops, and forestry compared with annual crops, as measured by share in gross value of production (CGIAR 1990).

The Role of Sustainable Agriculture in Reducing
Poverty and Improving Natural Resource Management

With 8 percent of the world's population, Latin America has 23 percent of the world's potentially arable land, 12 percent of its cultivated land, 46 percent of its tropical forests, and 31 percent of its fresh water. In addition, the Amazon contains about half the world's species of plant and animals. Despite its significant natural endowments, the potential of agriculture and natural resources to contribute to the social progress of the region has gone unrecognized by the World Bank and other international organizations. This is due in part to the fact that the agricultural sector in the LAC region is smaller than in either Africa or Asia in terms of both contribution to GDP (a declining 10 percent, compared with 25 percent in East Asia and 30 percent in Africa and South Asia) and rural population (only about 30 percent of the total population, compared with 50 percent in East Asia and 70 percent in Africa and South Asia). However, as discussed earlier, the aggregate picture is misleading, because important and alarming details go unseen.

The diversity and abundance of Latin America's agriculture and natural resources provide the region with an enormous comparative advantage with which to compete on world markets and thereby generate broad-based growth throughout the economy for urban and rural people alike. At the same time, international cooperation agencies and their governmental and nongovernmental partners cannot afford to neglect the natural resource management problems of traditional agricultural areas, where the largest concentrations of rural poverty exist. For most farmers living in these areas, agriculture is not just one of a number of ways to earn a living; it is the principal means of livelihood. Agriculture plays a key role in ensuring food security, and if it is neglected, this will irreversibly degrade the crucial natural resources that these populations depend on and perpetuate their poverty.

Characterization of Relationships Affecting
Natural Resource Degradation and
Sustainability in Agricultural Areas

The maintenance of the sustainability of land for agricultural use is a long-run issue of extreme importance for Latin American countries. The linkages between environment, poverty, and agricultural activities must be understood to guide future action. Agricultural activities, whether intensive or extensive, affect the natural environment. Environmental degradation, in turn, can compromise current agricultural productivity, undermine future production, and perpetuate poverty. Thus, while poverty accelerates environmental degradation, environmental degradation increases poverty.

Figure 1.1 indicates a series of possible relationships affecting natural resource degradation and sustainability in the region's agricultural areas. In using the framework, one should not focus on the inevitability or deterministic nature of the relationships leading to natural resource degradation. There are, of course, multiple trajectories for farmers facing a sustainability crisis that can result in better forms of natural resource management. However, many areas of traditional agriculture in Latin America are caught up in this downward spiral of economic and social disintegration and environmental degradation.

Rural Dualism: Traditional Low-Potential Versus Modern High-Potential Commercially Oriented Agriculture

Perhaps the most significant characteristic of the diversity in agricultural landscapes in many Latin American countries is that of rural dualism. This pattern is characterized by the coexistence of a relatively prosperous commercially oriented agricultural sector based on large-scale farms and an impoverished, traditional, often indigenous small-scale agriculture.[1] Traditional farms are not only small and unable to support a family's increasing consumption and production demands; they are also generally located in densely populated marginal areas on land with inherently low productive capacity characterized by poor or severely degraded soil, steep terrain, unfavorable climate or a combination of these factors.

Population growth also tends to be high in these areas, and despite high levels of out-migration, the size of local populations seems destined to increase for some time to come. In addition, off-farm employment and income opportunities are limited in traditional risk-prone areas, primarily because these activities depend on local demand for nonfood goods and services, which is low because local incomes from agriculture are low. Furthermore, degradation of the resource base limits opportunities for on-farm investment, and it undermines incomes and assets in ways that make sustained growth difficult. The upshot of this situation is an increasing number of poor people living in these areas, with a natural resource base that is further degraded by their own desperate quest for subsistence. Worst affected by this situation are the most vulnerable population groups—the poor and indigenous communities (Wiens 1994).[2]

There are also striking differences between the services provided by private and public agencies to commercial and traditional agriculture. In general, conventional public-sector approaches to agricultural technology development have had difficulty coping with the wide range of agroecological and socioeconomic conditions and institutional constraints (for example, weak markets, insecure land tenure, poor access to capital and knowledge)

Figure 1.1   Characterization of Relationships Affecting Natural Resource Management and Sustainability in Risk-Prone Environments

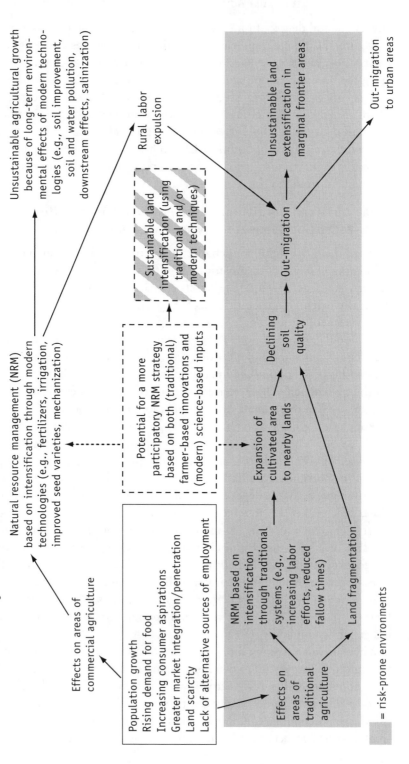

found in risk-prone areas. This situation has largely determined the menu of farming methods and technological options available to smallholders in these areas. The latter risk further neglect unless there is a rethinking of where public expenditure for research on technology development and dissemination should be directed and how it should be managed.

Survival Strategies and Linkages

Important linkages exist between poverty, population pressure, agricultural intensification, and natural resource degradation in traditional risk-prone areas. Poverty, combined with population pressures, land constraints, and lack of appropriate production technology to intensify agriculture, is a major source of environmental degradation in risk-prone areas in the LAC region. The effects of these factors on areas of traditional agriculture via increased food demands can be met by an increase in the land area in use, an increase in the intensity of use of existing land, or both.

The literature suggests that families historically tend to exhaust their economic responses before their demographic responses because of the greater psychic disutility of the latter (that is, expansion of the agricultural frontier or out-migration) (Bilsborrow 1992). The family first attempts to adapt by reducing leisure time and increasing labor effort and by changing technology—to the extent that alternative technologies are known and available.[3] However, when more land is available locally, the more likely response is land extensification. Land extensification can be achieved by (1) clearing more of the farmer's own land or, when such lands are no longer available and technological adaptations do not exist, appropriating nearby lands without migrating away from the family plot; or (2) migrating to other areas where trees and brush are cleared and people begin new cultivation. Depending on the location, this may involve using steep highland slopes, clearing brush in semiarid regions, or cutting down trees in commonly held forested areas. Unfortunately, as populations grow and there is little remaining unused land with high agricultural potential, the land extensification process becomes associated with the exploitation of marginal lands (including lowland tropical rain forests) in the absence of environmental policy controls or remedial measures (Bilsborrow 1992).

Until recently, geographic expansion of agriculture was a common option for many, if not most, Latin American countries. For example, much of the agricultural land in southern Brazil was still covered with natural vegetation well after the turn of the twentieth century. Under these conditions, yield-enhancing technologies were not critical, because expanding demands for crops and livestock could be satisfied by simply bringing more land into production (Southgate 1992). As Hayami and Ruttan (1985) point out, efficient agricultural development should begin with outward shifts in agriculture's extensive margin. Investments to improve crop and livestock yields (through changes in agricultural technology) are necessary once opportunities for geographic expansion start to be foreclosed (Hochman and Zilberman 1986). In general, development of North American agriculture has been consistent with this pattern.

Although unevenly distributed, current prospects for frontier expansion of agriculture in Latin America are becoming limited. The frontier is rapidly closing in areas such as southern Mexico, eastern Paraguay, northern Guatemala, Panama, Colombia, Ecuador, and other countries and regions with forest margins. The remaining frontier is now mainly in the Amazon. Nevertheless, in the Amazon region, as in many other marginal frontier areas around the world, virtually all the soils that are relatively suitable for crop and livestock production have been occupied by farmers and ranchers, and yields have tended to be modest at best on newly cleared land (Fearnside 1986). It is remarkable that during the past twenty-five years, every country in the LAC region except Argentina has lost more than 5 percent of its forests. Similarly, almost 50 percent of the Central American forests have been cleared since 1960.

The rural poor also tend to lose their capacity to support themselves sustainably when their access to resources is diminished or available resources are reduced. They may lose traditional access to resources if they are displaced by population pressure that reduces their access to land; by misappropriation of common resources by other claimants; by events such as wars, social strife, and natural disasters; and by development activities such as the construction of dams, the establishment of forest plantations, and the creation of wildlife preserves that take land out of use by the poor. In response, the poor may be forced to migrate to urban areas or ecologically marginal lands. Those forced to the latter may move higher and higher up hillsides or cut down forests for agricultural land, displacing indigenous inhabitants of these areas who may have developed sustainable resource-use techniques.[4]

Moreover, since agricultural expansion in the humid Latin American tropics is carried out largely by those displaced from traditional areas by poverty and land scarcity (and by other social and political pressures), the new, fragile lands are managed by those with the fewest resources to devote to their management. Under such circumstances, the need for technological innovation in agriculture is greater, but the means with which to innovate often appear to be lacking. Hence, declining productivity is compensated for primarily by bringing even more land (which is usually of marginal quality) into production rather than by maintaining or increasing productivity over the long term. Degradation of the natural resource base in these areas is thus an inevitable consequence, at least in the short term.

Although Figure 1.1 and the preceding discussion suggest some of the possible relationships affecting natural resource degradation and sustainability in the region's traditional agricultural areas, the conditioning influence of the particular context within which these relationships occur must also be taken into account. That is, consideration must be given to the extent to which extensification or intensification of agriculture takes place in a specific country or area during a given period, the forms in which it occurs, and the prevailing natural resource endowments (including land, geography, climate, rainfall) and institutional and attitudinal factors (which are derived ultimately from the country's history and contemporary government policies).[5]

Agricultural Intensification in High-Potential Agricultural Areas

Agricultural intensification in high-potential commercial agricultural areas has been based mainly on a combination of inputs such as fertilizers and pesticides, plant-breeding technology to develop high-yielding and pest- and disease-resistant varieties, and irrigation technology. Notwithstanding the substantial contributions to the regional increase in food production attributable to these modern technologies, there are serious concerns that the agricultural intensification process accruing in these areas is leading to severe degradation of the natural resource base (Pinstrup-Andersen and Pandya-Lorch 1995). Increased destruction of habitat and wildlife, soil erosion, desertification, waterlogging and salinization of soil, contamination of surface and groundwater, extinction of species, resurgence of pests, and other forms of natural resource degradation in recent decades have been attributed to mismanagement of external inputs and modern green revolution technologies.

Despite the wide array of life-forms and ecosystems in Latin America, green revolution technologies were introduced primarily in the humid pampa of South America, the irrigated areas of Mexico, the Caribbean islands, the Pacific coast, and some tropical areas. Between 1961 and 1990, 71 percent of the increased production in the LAC region resulted from increased yields obtained mainly from these areas, and the remaining 29 percent was obtained from area expansion (World Bank 1992). Nevertheless, past and present research and extension efforts have not produced significant new technologies for the lower-potential and more marginal areas, mainly in Central America and the Andean region, where most traditional agriculture and rural poverty are concentrated. Extension, like research activities, has suffered from budget cuts and salary caps, and the traditional areas have not benefited—as has commercial agriculture—from the development of private suppliers of technical and farm management information and counsel.

Government policy has tended to ignore resource-poor farmers in risk-prone environments, except as pools of labor or target zones for welfare transfers. Premature shedding of labor from the commercial farm sector and failure to make alternative arrangements for the rural labor force in smallholder sectors cause the latter to sink deeper into poverty or to become slum dwellers in cities.

Many areas of traditional agriculture in risk-prone environments can be found somewhere in the midst of this type of downward spiral of natural resource degradation and economic and social disintegration. The usual do-nothing policy response regarding traditional areas has encouraged continued insecurity in such areas and continual migration of the destitute to urban and marginal rural areas, both of which entail huge social costs (Wiens 1994).

Agriculture, Technology, and Sustainable Livelihoods

Modern Green Revolution Technologies and Mismanagement of the Agricultural Intensification Process

The introduction of high-input agriculture under the green revolution initiative channeled scarce investment resources into capital-intensive agriculture, which became dependent on imported machinery, equipment, hybrid seeds, agrochemical inputs, and irrigation. The development of modern seed varieties of food crops such as maize, wheat, and rice and other commercial crops played a key role in relation to the other components of the farming system (fertilizer, irrigation, plant protection, and mechanization). By and large, it is undisputed that under otherwise unchanged conditions, Asia could not have escaped widespread famine without the new seed varieties. The gains in productivity, total grain production, and associated income gains to producers and consumers resulting from the green revolution have been and continue to be enormous (Alexandratos 1988; Hazell and Rama-samy 1989; Pinstrup-Andersen and Pandya-Lorch 1995).

Still, the new capital-intensive technologies worked well where ecological conditions were relatively uniform (such as in irrigated areas) and where delivery, extension, marketing, and transport services were efficient. This made the new technologies more attractive in areas that could produce a high return to the required investments. However, since the early productivity successes of the green revolution, it has not been possible to even remotely approach the yield increases recorded during the 1960s and 1970s. Failures to maintain past levels of investment in agricultural research, technology, and irrigation, along with falling real agricultural prices and associated decreases in the use of fertilizers, are reportedly contributing to falling yield growth rates for the major cereal crops of rice and wheat (Rosegrant and Svendsen 1993).

More importantly, there is increasing anxiety that the goal of meeting current and future food needs may be in conflict with the goal of protecting the productive capacity of the natural resource base. The experience with the green revolution in several parts of the world has shown that when cultivation is intensified with the aid of fertilizers and pesticides, new seed varieties, controlled irrigation, and mechanization, the benefits of short-term yield increases must be weighed against the costs and ecological risks. Excessive use of fertilizers and plant protection agents (such as pesticides and fungicides) and overintensive mechanization not only are costly but also have caused unwanted effects on the natural environment and human health (Pinstrup-Andersen and Pandya-Lorch 1995).

Poor management of irrigation has led to considerable degradation of the natural resource base in many green revolution areas. About 20 to 30 million hectares worldwide (or 10 percent of all irrigated land) suffer from severe salinity, which, if not treated, can ultimately destroy the land; another 60 to 80 million hectares experience waterlogging and salt buildup. In Mexico alone, more than 50,000 hectares have been abandoned due to salinity. Salinization removes about 1 million tons a year from Mexico's grain harvest. The salinity problem is growing, with as many as 1.5 million hectares lost each year around the world, equivalent to about half of the new land brought into irrigation. Besides environmental consequences, the increased prevalence of malaria and schistosomiasis in many irrigated areas has been noted for some time (FAO 1993; WHO 1980, 1983).

Inappropriate or excessive pesticide use has significant environmental and social consequences. Repeated application of broad-spectrum pesticides has led to the buildup of resistance in target pest species (FAO 1989; Hansen 1987; Weber 1992). Besides the resurgence of known pests, secondary pests are emerging from nontarget species—those that were not originally pests themselves but whose natural enemies were unwittingly destroyed by repeated applications of pesticides. For example, in Mexico, pesticide treatment to control the cotton boll weevil and the pink bollworm also destroyed the natural enemies of a beneficial insect and decimated the cotton industry—acreage under cotton fell from 300,000 to 500 hectares during the 1960s (Pinstrup-Andersen and Pandya-Lorch 1995; Hansen 1987).

Pesticide poisoning is also a serious problem where widespread and indiscriminate application contaminates soil and pollutes water through runoff and leaching into groundwater. In Honduras, the greatest threat to the Gulf of Fonseca is water pollution caused by the misuse of pesticides (DeWalt, Vergne, and Hardin 1993). Worldwide, some 20,000 deaths and 1 million illnesses each year are attributable to pesticide poisoning, and most of these cases occur in developing countries (Conway and Pretty 1991). Improper pesticide use impairs farmers' health, which, in turn, affects farm household productivity. Exposure to pesticides can lead to cardiopulmonary disorders, neurological and hematological symptoms, and skin diseases (Rola and Pingali 1993). Effects of pesticide exposure can also be passed on to infants through poisoned breast milk. In cotton-growing regions of Nicaragua, breast milk samples from women contained some of the highest levels of DDT ever measured in human beings (World Bank 1992).

Still, agricultural intensification, if managed properly, need not degrade the environment. There is strong evidence that agricultural productivity can be significantly increased by combining traditionally practiced farming systems and modern agricultural techniques.[6] Alternative technologies and farming practices that are potentially profitable to farmers include appropriate crop rotations, mixed farming systems with crops and livestock, agroforestry, biological pest control, disease- and pest-resistant varieties, balanced application and correct timing and placement of fertilizer, selective use of pesticides, and minimum or zero tillage.

Before discussing the important role that these systems can play in intensifying agriculture and reducing poverty in risk-prone areas, we would like to examine the sociopolitical impact of modern technologies and the alleged effects on resource-poor farmers. At the heart of this debate lies the recognition by both critics and supporters of the new technologies that the question is not *whether* but *how* to intensify agricultural production.

Many analysts agree that the most serious environmental problem in risk-prone areas is not inappropriate technological change but the many millions of people who live in absolute poverty. Poverty is a significant force behind the lack of incentives for smallholders to innovate with or apply appropriate technologies, and it clearly restricts access to other critical inputs required for agricultural intensification. We believe that if poverty is not systematically tackled in risk-prone areas, traditional agriculture will become even more marginalized, with further natural resource degradation and increasing out-migration of resource-poor farmers to urban and frontier areas. Economic viability needs to be guaranteed to ensure sustainable livelihoods for rural families—a challenge to modern technologies and agricultural policies.

The Achievements and Limitations of Modern Technologies

The sociopolitical ramifications and ecological impact of the green revolution in developing countries are complex and controversial. There is ongoing debate about whether there was a causal relationship between the technologies of the green revolution and the incidence of rural poverty (see Mellor and Desai 1985; Rhoades 1988; Thrupp 1989; DeWalt 1995).

Unfortunately, clear empirical information on the dimensions, causes, and even locations of problems is lacking in many developing countries. The discussion by Reijntjes, Haverkort, and Waters-Bayer (1992) of green revolution technologies and modern agricultural research offers several interesting explanations of their bias toward high-potential areas, export crops, and better-off farmers:

1. The emphasis of green revolution technologies for agricultural development has been on maximizing the production of particular commodities, not total farm production. Given the focus on single crops, the long-term effects on soil fertility and the regenerative capacity of natural vegetation and fauna have not usually been given sufficient consideration. This has also hindered the study and improvement of positive interactions between different plants and animals that, in addition to providing farmers with livelihoods, can contribute to the continuity and stability of farming. The promotion of modern varieties has also led to the disappearance of many indigenous varieties. This may spell disaster for farmers who have to produce their crops with low external inputs under highly variable and risk-prone conditions.

2. The recent stagnation in the production increase has raised doubts as to whether the long-term productivity of modern technologies is secure. Modern seed varieties are essentially high-response varieties, bred to respond to high doses of chemical fertilizers. If they are sown under conditions of high nutrient and water supply and adequate pest control, modern varieties and hybrids can be and have been high yielding. However, when these conditions cannot be guaranteed, such as in risk-prone environments, risks of yield losses may be higher than with local varieties. According to Dover and Talbot (1987), as much as 80 percent of agricultural land today is farmed with little or no use of modern irrigation, agrochemicals, machinery, or improved seeds.

3. Farmers who are given access to credit may be required to engage in high capital investments and production methods that demand high levels of external inputs. However, when purchased inputs are subsidized by the government or a development project, their use is feasible only for a limited time. As soon as subsidies are removed and farmers are forced to abandon the inputs, it is unlikely that they will be able to adjust other aspects of their farming systems (for example, reduced diversity of crop and livestock species or increased nutritional dependence on crops such as maize, which require high fertilizer inputs) back to the original conditions without adverse consequences.[7]

4. The conventional top-down approach of technology development within agricultural research institutions has given scientists little opportunity to become well acquainted with the wealth of local genetic resources and traditional knowledge about ecologically oriented husbandry and local alternatives to purchased inputs. Similarly, the production conditions of experiment stations seldom resemble those of farmers. Consequently,

technology tested in stations has not worked under farmers' conditions, and good qualities of local varieties, which are adapted to local conditions, are not recognized under station conditions. The products delivered for extension have tended to be incomplete and designed without sufficient regard for household issues such as risk spreading, labor allocation among other existing crops, affordability of modern inputs, and other crucial socioeconomic aspects. Also, until recently, conventional agricultural research has given little attention to important questions of women's influence on decision making and labor allocation in the design of new technologies and extension systems.

The implications of modern technologies for income distribution in rural communities have also been the subject of much debate. Since the new technologies save land by permitting more intensive use of labor and external inputs, they might be expected to contribute to a more favorable income distribution among farm households. However, modern technologies have often been criticized for benefiting landlords at the expense of tenants and laborers in high-potential areas, on the grounds that land rents have increased where modern crop varieties have been introduced. Nevertheless, evidence from some parts of the developing world has shown that the adoption of modern varieties resulted in increases in labor demand, even in areas where it was accompanied by concurrent progress in mechanization. Despite this evidence, analysts have documented growing inequalities emerging in many green revolution areas. The extent to which these inequalities can be attributed to the new technology itself or to insufficient progress in its development and diffusion is, however, unclear.

Three key arguments are presented by analysts in challenging this critique of the green revolution:

1. Technology is not responsible for sociopolitical inequalities (Leisenger 1995). Sociopolitical contexts determine whether a technology leads to inequalities or not.

2. Technology innovation at previous speeds is no longer sufficient for contemporary food demand. Developing countries need a combination of traditional and modern technologies to ensure sufficient agricultural growth.

3. High labor demands, and thus food costs, would have increased poverty and ecological degradation, especially in resource-poor areas without modern technology (Hayami and Ruttan 1987). Institutional reform toward efficiency of resource use is encouraged only by the dependence of modern technology on efficient markets and input availability.

Traditional Farming and Innovation

Are advances in traditional or indigenous technology sufficient to sustain rising levels of per capita income and consumption in risk-prone areas? Or, at least, what role should indigenous technology and management systems play in helping to attain such goals? Many traditional farming systems were sustainable for centuries due to their ability to maintain a stable level of production (TAC/CGIAR 1988, 1993). A wide range of different farming and animal husbandry systems has evolved, each developed to adapt to the local ecological conditions and inextricably linked with the local culture. Traditional systems have changed quickly, however, particularly over the last few decades, as a result of increasing population pressure, market integration, indiscriminate promotion of modern inputs, and financial constraints. The impact of population pressure on inducing indigenous improvements in agricultural technology in contemporary farmer societies has been documented by an increasing literature led by Esther Boserup. Her insistence on the importance of population growth in inducing the development of intensive systems of agricultural production challenges the view that agricultural technology in traditional farming communities was essentially static.

Examples of Traditional Methods of Natural Resource Management

There are no technical blueprints for exploring the possible contributions of indigenous practices to new technology developments. Nevertheless, techniques that hold promise and that are most likely to be applicable in the various contexts are those involving careful conservation of soil and water, use of complementary or symbiotic genetic resources (intercropping, integrating trees and animals), advantageous use of nitrogen fixation, and complementary and efficient use of external nutrient inputs (natural or artificial). Many traditional practices, not all of which are known to formal agricultural science, represent at least the seeds of promising new technologies based on composting, green manuring, mulching, multiple cropping, contour farming with hedges, water and nutrient harvesting, and controlling pests. If these indigenous practices are understood well in formal scientific terms, it may be possible to improve them, for example, by careful use of external inputs and modern technologies. Also, many indigenous, sometimes unconventional crop and animal species and local varieties and breeds may have great potential for new technologies in risk-prone areas.

A growing number of publications are now appearing about indigenous knowledge systems and the farming systems based on them (see Brokensha, Warren, and Werner 1980; Biggs and Clay 1981; Rhoades 1984; Richards 1985; Marten 1986; Wilken 1987; Warren and Cashman 1989; Chambers, Pacey, and Thrupp 1989; Warren 1991; Reijntjes, Haverkort, and Waters-Bayer 1992; Pretty 1994; Scoones and Thompson 1994). Case studies of successful experiences include experimentation or innovation in food and tree crops, irrigation and other water harvesting techniques, gardening, seed distribution, field and seed preparation, fertilization, livestock nutrition, rodent and weed control, natural resource management, food storage, food processing, and market products and outlets among many others.

A common theme running through most of this work is that traditional farming systems are in constant change. An attempt is continually being made to adapt to the new conditions imposed by population changes, greater aspirations, market integration and so forth. These adaptations have not, however, always been adequate, and entire cultures have disintegrated as a result. Many indigenous practices that in the past sustained human populations for centuries have become obsolete as conditions have changed. For example, several forms of shifting cultivation have proved to be unsustainable under increased population pressure, and as a result, they cannot be maintained.

Yet there are literally hundreds of studies that have recorded the importance of traditional knowledge systems in successfully conserving natural resources in many countries.[8] A wealth of gray literature (unpublished reports and articles in newsletters, circulars, dissemination notes, and project reports) also reveals the experiences of farmers and supportive scientists in a wide range of nonconventional forms of agriculture such as low-external-input agriculture, organic farming, biodynamic farming, and permaculture. NGOs have been particularly active in disseminating this type of work.

Participatory Technology Development

An important contribution made by the nongovernmental sector to these positive developments has been the promotion of methodologies for combining local farmers' knowledge and skills with those of external agents to develop and/or adapt appropriate farming techniques—participatory technology development. Farmers work with professionals from outside their community, such as researchers and extensionists, in identifying, generating, testing, and applying new technologies. Participatory technology development seeks to strengthen the existing experimental capacity of farmers and encourage continuation of the innovation process under local control (Haverkort et al. 1988). Numerous case studies illustrate the range of initiatives in participatory technology development,[9] such as the successful participation of local farmers in problem identification, including substantial reorientation of initial objectives defined by researchers.

From the perspective of research and development institutions of the National Agricultural Research System (NARS) and the Consultative Groups on International Agricultural Research (CGIARs), of particular importance for technology development is the capacity of farmers to understand the local biophysical and cultural environment and to predict and explain the outcome of experiments under local conditions (Ashby 1987;

Ashby et al. 1990, 1995). Wanting to support technology that is appropriate for smallholders, these researchers present cases in which farmers' evaluation of technology has provided researchers with new insights and in which farmer-to-farmer dissemination has been successful. When researchers already have good knowledge about a technology, a common and cost-effective technique is to offer farmers several technology options related to the problem or opportunity at hand and leave it to them to experiment in an ad hoc fashion (for related experience and discussion, see Anderson 1996).

As they become familiar with new technology, farmers are likely to change other components of their farming system to exploit the advantages it offers. Such changes can be complex and variable over time and space, so researchers have little prospect of predicting them on the basis of their own trials (Ashby et al. 1995). The evaluation criteria used by farmers and observed by researchers can then be fed into the next round of technology development for release to farmers (Farrington and Martin 1988). The search for new participatory methods has also led to efforts to meet both researchers' and farmers' requirements in a single set of trials, usually via interfarm instead of intrafarm replication. These trials have been particularly useful in accelerating the release of new genetic material (Maurya, Bottrall, and Farrington 1988), but in some cases they have been costly and have led to failure, prompting a move back to on-station trials.

Regarding the types of technology being developed, reports of field experiences in the selection of genetic material outnumber reports of any other application. Important but isolated examples have been recorded in the management of soil, water and forest resources, crops, and storage facilities. Examples of crop protection, fertilization, farming equipment, and livestock research are less numerous. The focus on genetic material perhaps highlights the area of greatest complementarity between researchers and farmers. The latter have a vast range of material on which to draw and have developed breeding methods exceeding in both scope and speed those available to the former. It is remarkable that more than 80 percent of the crops cultivated in developing countries are planted with seeds saved from the preceding season and from informal farmer-to-farmer seed systems (for example, informal potato seed systems in Peru). These informal systems are strong even in countries that have relatively advanced seed industries (Cortes 1995; Jaffee and Srivastava 1992; Wiggins and Cromwell 1995).

Potential for Mediation of Interests

Most of the current experience in participatory technology development is less than clear about the actual success of participation in practical terms, particularly regarding the role of extension once researchers and farmers have been drawn closer together in a participatory approach. Similarly, the role of local organizations and of nonindigenous NGOs in articulating client demand for and in mediating participatory inputs into agricultural research has received little attention, in spite of its considerable potential.

There are numerous case studies of projects using innovative methods at the outset, but substantive evaluation of this experience is still scarce, and the time- and cost-related effectiveness of participatory methods is not yet well documented.[10] Also, although field experiences all recognize that participatory approaches can lead to greater cost-effectiveness, not only in problem-focused but also in commodity and factor research, precisely how the results of the participatory methods influence the agenda for research in these areas is rarely illustrated from empirical evidence.

Concerning the institutional framework, practically all participatory technology development experiences have been undertaken outside national agricultural research programs. Numerous examples have had a continuing institutional base through the CGIAR centers, NGOs, and universities. However, many more have emerged from specific research projects of limited duration with no apparent commitment to their eventual incorporation into any institutional framework. It is clear that many more cases are needed of attempts to incorporate participatory technology development into national agricultural research programs.

As a result of decreasing levels of both government and international donor funding, NGOs are moving to fill the vacuum left by the decline of public-sector extension services. They are assuming greater responsibility in identifying and distributing required inputs, such as suitable seed, and in providing technical support services in risk-prone areas (Farrington and Bebbington 1994; see also Chapter 2 of this book). Some funding agencies have supported collaboration between government and nongovernmental entities, utilizing the latter's capacity as brokers between farmers and research services. The capabilities that NGOs bring to bear are derived from their close knowledge of the needs and opportunities of the rural poor in relation to agricultural change, not merely in the sense of crop or animal technology but in the wider context of innovation. The types of institutional forms and changes required to ensure full incorporation of participatory methods of technology generation into the work and policies of universities and research and extension and public-sector agencies, however, are as yet insufficiently documented.

Farmer Participation and Traditional Knowledge: Development and Dissemination

In their constant struggle to sustain their agricultural livelihoods, resource-poor households and communities in risk-prone environments have developed innumerable ways of obtaining food and fiber from the natural environment.

In many risk-prone areas throughout the developing world, long-term concern for the sustainability of the natural resource base has been an important traditional management objective, often institutionalized in local regulations and cultural norms (Warren 1995; Rhoades 1988). However, when poverty has become so extreme that only day-to-day survival can be achieved, traditions of natural resource conservation may disappear. At issue here is that there is no margin for reconciliation with ecological sustainability at the present levels of poverty found in risk-prone areas. Thus, it is naive to argue that farmers living in areas of traditional agriculture would not be interested in incorporating modern agricultural technologies into their farming systems, if they had the means and opportunities to do so.

To ensure the continuity of their livelihoods, farmers must be able to adjust to change driven by increasing demands (for example, population growth, greater market integration, desire for more consumer goods). Vital to such adaptability is the capacity to manage farm development, to choose appropriate combinations of genetic resources and inputs, and to develop new technologies and fit innovations into the farm system to raise output in subsistence food crops and commercial crop production. This legitimate need for a mix of both modern and traditional technologies does not obviate the fact that in poor areas, especially where farmers depend mainly on local resources, modern technologies may not be the first or the only option to improve agriculture. Strengthening farmers' and development agents' understanding of the ecological principles underlying contemporary resource use and farming and adding to their knowledge of the available technical options are important steps in the process of strengthening farmers' capacity to develop and manage technology for sustainable development (Smith 1995). This also implies that the solutions to farmers' problems will be as diverse, complex, and site-specific as their farming systems, but they will be more "owned" by farmers.

The crucial issue is not, therefore, whether one type of technology should replace another, but how—in terms of methods and institutions—the most relevant aspects of each can be brought to bear on the natural resource management issues of a particular area (DeWalt 1995). The path outlined might look as follows:

- *Understanding and knowledge development.* Agricultural sustainability in risk-prone areas requires an understanding of the diverse and complex environments in which resource-poor farmers operate, so that developments in technology can be tailored to suit their circumstances, building wherever possible on farmers' indigenous technical knowledge.

- *Development and dissemination of combined methods.* This might involve combining chemical and organic fertilizers, appropriate forms of green manuring, or the integration of new crops. Knowledge about these technical options would be spread, and the forces of farmers, field-workers, and scientists could be combined to discover their opportunities and limitations.

■ *Democratization and decentralization.* Democratization of political processes, decentralization of unwieldy institutional structures, and effective application of the principle of subsidiarity are needed for greater government accountability and farmer participation in developing new approaches to technology development, support services (such as local gene banks and seed multiplication), and legislative provisions (for example, government certification and release of new varieties).

■ *Assured farmer participation in research.* The diverse criteria by which farmers assess any increased production and welfare benefits offered by technological change are central to the technology development process. Their collaboration—in knowledge, skills, and time—is required to enable scientists at national and international research centers, facing restricted budgets for work with smallholders, to undertake their work in such a way as to ensure appropriate technology development.

Building Blocks for a New Strategy Toward Traditional Risk-Prone Areas

By way of conclusion, then, we draw together the lessons to be learned from the developments outlined earlier, making recommendations that are general in nature but that are already being undertaken by different actors in the Latin American and Caribbean region (as several chapters in this book indicate).

In pursuing efforts to reduce poverty, increase agricultural productivity, and improve management of natural resources, policymakers, governments, and donors are confronted with the fundamental question of whether to focus their attention on high-potential areas or on low-potential, risk-prone areas. Many high-potential areas are now degraded or suffer from environmental stress, and there is doubt whether high-potential areas will have the capacity to meet future food needs in a sustainable manner. Yet a large proportion of the region's poor live in these highly vulnerable areas, and the risk of destruction of the natural resource base as a result of their survival strategies in the absence of external assistance is high.

But difficult trade-offs arise in investing in the development of risk-prone areas—at least in the short term—and these need to be recognized by development agencies. Although agricultural growth can usually be best achieved through investments in the highest-potential regions, rural poverty and resource degradation problems are often located in low-potential regions. Public investment in traditional agricultural areas may have high opportunity costs relative to the areas of commercial agriculture, where most agricultural growth has occurred. However, efficiency trade-offs may be justified in terms of the greater impact on poverty and environmental degradation that can be achieved in traditional areas. Therefore, although alleviating poverty and improving natural resource management are not necessarily linked to promoting significant agricultural growth in traditional areas in the short term, in the long run, poverty alleviation is necessary for achieving sustainability in these areas.

A New Paradigm of Sustainable Development in Risk-Prone Areas

The discussion in this chapter suggests that a new paradigm needs to be developed that integrates risk-prone areas into rural growth (Delgado 1995). Although there is significant agreement on what is needed to boost production in the highest-potential areas, there is little consensus on the other 80 percent of cropped area worldwide. Experience suggests that attempts to make lower-potential zones do the same things as higher-potential ones are problematic. Some argue that the best policy to pursue in low-potential, vulnerable areas is to relieve population pressures by encouraging massive out-migration. Although, in the long run, out-migration may be an answer for some traditional areas, merely transferring poverty and population pressures to urban areas and forest margins cannot be an acceptable solution to rural problems.

High-potential areas that have been degraded must also be rehabilitated and their productivity restored to the extent feasible. Although more attention must be paid to areas with fragile ecosystems and large numbers of poor people, further improvements in the agricultural productivity of high-potential areas must be pursued to meet the needs of the rapidly increasing urban populations. In a sense, the successes of the green revolution have bought us time to adjust the intensification approach in these areas to ensure that it is sustainable in the long term. In addition, policymakers might explore how to assist the lower-potential areas to benefit symbiotically from demand pull in higher-potential areas. This symbiosis could be obtained through agricultural diversification in the lower-potential areas into items consumed in the adjacent higher-potential areas (such as staples or niche market foods).

Intersectoral Linkages

It is evident from the preceding discussion that the focus of agricultural development must be broadened to include intersectoral linkages that affect employment generation, poverty reduction, and related issues of population growth and expulsion and attraction forces driving rural migration to the forest frontier and urban areas. Generally, the dimensions of how investment in agricultural research and extension, health, education, and urban services may influence decisions on migration and resource use in a particular region are paid no attention in the investment allocation procedure by international development agencies and governments.

Infrastructure development, growth of the nonfarm economy, and decentralization of financial and administrative decisions are key aspects to be considered in the new paradigm. The relative lack of public- and private-sector investment in infrastructure and human resources in many traditional areas of the region has inhibited the ability of agriculture and natural resources to realize their full development potential. To reverse this neglect, governments in the region will need to allocate the resources necessary to develop and maintain productive infrastructure and services in resource-poor areas, including crop storage and processing facilities, irrigation, roads, telecommunications, financial services, and supply channels for inputs.

In many countries, the potential multiplier benefits of agriculture are constrained by investment codes and related legislation that discriminate against small, rural nonfarm firms. These policies need to be redressed. The rural nonfarm economy (as well as rural-urban linkages generally) can be strengthened by removing institutional barriers to the creation and expansion of small-scale credit and saving institutions and making them available to small traders and to transport and processing enterprises.

In fact, agricultural growth creates powerful multiplier effects on the rural nonfarm economy that enable farm households in degraded lands to mobilize capital and labor for farm investment and rehabilitation, income diversification, evolution of property rights, and infrastructure investments, leading to additional employment and earning opportunities for the rural poor (Scherr and Hazell 1993). In addition, off-farm employment (including income-earning activities in rural industry, services, and marketing) is an important component of the survival strategies of the rural poor and should be an important focus of antipoverty efforts. Given its importance for overall sustainability, it is essential that the policy framework for agricultural development be broadened to incorporate the relationship between geographic and intersectoral allocation of public expenditures influencing employment, poverty, and migration.

Governments in the region must also improve the quality of education, health care, and sanitation in traditional areas. Productive and social sectors are synergistic, not competitive. Just as investment in health and education can increase productivity and help achieve economic goals in traditional areas, investment in the development of agriculture and natural resources can help achieve social goals. Governments need to revitalize local governments in traditional areas and create an enabling environment for local institutions to identify, develop, and maintain new infrastructure and services. To improve efficiency, governments need to recover costs through user fees; identify and select projects, with the full involvement of farmer communities, based on careful evaluation of potential demand for services; and involve NGOs and private contractors in executing projects.

Technology Development

Even with the benefits to be gained from economic liberalization, one must be aware that the increasing emphasis on structural adjustment is, in the short run, likely to increase the number of vulnerable and at-risk people in the region. As the state withdraws from organizing rural economic life, a vacuum is left that is not fully filled by the private sector. Hence, interactions between public-sector agricultural research systems, farmers, private companies that conduct research, private enterprises in food processing and distribution, and NGOs need to be strengthened to ensure the relevance of research and the appropriate distribution of responsibilities. This will make government policies, research priorities, and public-sector spending more effective and more responsive to the resource constraints and local ecological conditions faced by farmers, while providing a better foundation for interaction between government and civil society.

To ensure participation in agricultural research and extension and technology development, more resources need to be allocated by governments and their donors to promote intensive consultations with prospective beneficiaries. Technology development must be not only on-farm and farmer managed but also participatory in order to draw on local knowledge and consider farmers' needs, opportunities, constraints, and aspirations (Farrington and Bebbington 1994; Bebbington et al. 1993; Bebbington 1995; Chambers and Pretty 1994). Thus, enhancing the role of farmers in local analysis, in the setting of priorities, in experimentation, and in other research and extension activities is an important element of a demand-responsive strategy for dealing with rural poverty and natural resource management problems in risk-prone areas.

Correction of Market and Policy Distortions Through Incentives and Regulatory Measures

Incentives and regulatory policies need to be strengthened to compensate for externalities related to natural resources. The nature of such policy measures will vary across countries and over time, but they are likely to include policy reforms dealing with water allocation mechanisms and watershed management, exploitation of lands and forests resulting from free access, determination and enforcement of property rights (including land, water, and forests), and correction of distortions in input and output markets, asset ownership, and other institutional and market distortions adverse to the poor.

Regulations are necessary where incentives are unlikely to achieve social objectives. However, regulations that contradict the survival strategies of resource-poor farmers or their communities' customs are unlikely to be successful, simply because they will be difficult or impossible to enforce in resource-poor areas. Finally, serious commitment to promoting equitable access to land, water, and capital in traditional areas is crucial for achieving sustainable development and poverty reduction in the region.[11]

Political Leadership and Empowerment of Farmers in Risk-Prone Areas

An important aspect in the development of traditional areas is the strengthening of social organizations, such as indigenous organizations and farmers' and women's groups. These groups have endured historical patterns of exclusion that persist today. The violence endemic to the region compounds these difficulties. Although in many cases this violence is born of historical social, economic, and political inequalities, in others, such as in the Andean region, it reflects the rising influence of those associated with the illegal drug trade. This influence has infiltrated the societies of almost every country in the region, distorting the economic and social development of rural areas and undermining the confidence of citizens in public officials and in the entire justice and political system (Garrett 1995).

After decades of advocacy and support for the smallholder sector by scholars, donors, NGOs, and private voluntary and human rights organizations, there is wide recognition of the need for farmer empowerment. As long as small farmers in risk-prone environments continue to be at the periphery of the political process, and as long as governments continue to favor the politically powerful urban population, their poverty-stricken and ecologically vulnerable areas will remain ignored, except as pools of labor or target zones for welfare transfers. Consistent with human rights and democratic governance, legally and institutionally resource-poor farmers need to be able to group together to ensure that they have their own political voice.

The weak legitimacy of many governments in the region also constrains those enlightened policymakers who desire to formulate and implement rural strategies that are capable of mobilizing the population in the interest of rural families' livelihoods. Hence, effective support for LAC countries that are moving toward policies conducive to sustainable agriculture and rural development (SARD) and that can serve as models to other countries in the region is just as urgent as reducing donor support for governments that neglect resource-poor rural people. It is, however, not easy to turn the lending tap on and off, and some areas are gray. But influential international development agencies such as the World Bank have been too timid, and sometimes inconsistent, in pursuing their overriding objective of poverty reduction and in explaining that to Latin American leaders. Improvement in income distribution, although officially recognized as an essential means of fostering economic development and socioeconomic stability, is still not operationally relevant to policy decisions in most countries in the region.

The difficult choices and decisions with a long-term view that addresses the costs and trade-offs of resolving these conflicts are still to be made. Nevertheless, at least for now, the World Bank is one of the few institutions that have the influence, experience, and resources to mediate for and carry through a consistent strategy for SARD in the risk-prone environments of Latin America.

Notes

The authors are social scientists at the World Bank, Washington, D.C. The findings, interpretations, and conclusions are the authors' own and should not be attributed to the World Bank, its executive board of directors, or any of its member countries.

1. Rural dualism has long been defined as "functional dualism" by DeJanvry (1981) and others. The presence of such rural dualism is recognized in two recent World Bank documents delineating regional strategies for the agricultural sector and rural poverty alleviation in Latin America (World Bank 1993; Wiens 1994), although its extent and different forms are not documented. Nevertheless, the former document recognizes that this phenomenon is "at the root of many economic and political problems, threatening social harmony and posing hard choices for agricultural strategy" (World Bank 1993, 3).

2. Wiens (1994) reports that there is also a close correlation between ethnicity and rural poverty in risk-prone areas. Indigenous people are usually landless farm laborers in commercial farming areas or smallholders in areas of marginal quality and weak markets. A recent World Bank study estimates the indigenous population of Latin America to be between 19 and 34 million people, with the great majority living in Bolivia, Ecuador, Guatemala, Mexico, and Peru (Psacharopoulos and Patrinos 1994). The same study calculates that 80 percent of the indigenous population is poor in relation to the lower estimate of 19 million, and over half is extremely poor.

3. Referring to the effects of population growth on agriculture and technological innovation, Boserup (1965, 1981) challenged the (Ricardian-Malthusian) assumption of constant technology, postulating that as rural land becomes scarce relative to population (because of population growth), land will be used more intensively to produce greater yields. She then proposed that "the growth of population is a major determinant of technological change in agriculture," which stems from the necessity to raise output per unit of area to offset the increase in labor requirements associated with more intensive land use (Boserup 1965, 56).

4. In the process of extensification and the resulting natural resource degradation, property rights to land, forests, and water have become particularly prone to externalities. Resources with open access are particularly easy to overexploit, because users may benefit without paying the costs associated with reduced

future productive capacity. In addition, indigenous institutions for managing common property resources are breaking down, partly through misguided efforts by government and international institutions to privatize common property without a thorough understanding of these common property rights. For example, it is inappropriate to privatize property rights where people depend on key, spatially concentrated resources or when it could lead to "parceling up" of resources.

5. For example, the greater the availability of potentially arable land resources, the more likely is extensification instead of intensification (Pingali and Binswanger 1987). The country's level of development, government policy priorities, and ability to finance roads and infrastructure to open up new lands are also important. Institutional factors and government policies regulating access to land (for example, land tenure, concentration of landholdings, privatization) and land use (for example, price, taxation, and import-export policies; agricultural research and extension services; credit availability) can play major roles as well.

6. Although the term "indigenous knowledge systems" has become standardized in the literature, in this chapter we use the terms "traditional," "local" and "indigenous" systems, which encompass the combination of knowledge, productive resources, input, and services applied systematically by rural peoples to produce desired outputs. They include both the physical forms of technology (tools, seeds and so forth) and the methods, practices, and strategies, including forms of social organization. Finally, rural peoples include not only native peoples but also peasant farmers, settlers, and other resource users.

7. Many governments have promoted fertilizer and pesticide use by subsidizing, explicitly or implicitly, their prices. Mechanisms such as access to foreign exchange on favorable terms, tax exemptions or reduced rates, easy credit, and sales below cost are used to promote pesticide use (Repetto 1985).

8. For example, there are studies illustrating the growing interest by developing countries and donor agencies in the role of indigenous knowledge in making development projects more effective and efficient (Franke and Chasin 1980; Bheenick 1989; Verhelst 1990; Barborak and Green 1987; Hoskins 1984; Wilcox 1991, 1995). Niamir (1990) compiled a list of pastoral projects and programs that have included indigenous knowledge components. Titilola (1990) and Dommen (1988, 1989) analyzed the costs and benefits of adding indigenous knowledge components to development projects. Others have addressed the potential role of indigenous knowledge in international agricultural research (Cashman 1989), in forest management (Poffenberger 1990), in gender issues and development (Norem, Yoder, and Martin 1989), in sustainable approaches to agriculture and development (Ascher and Healy 1990; Jodha 1990; Warren 1991; Warren and Cashman 1989), and in the agricultural research and extension process (Butler and Waud 1990; Cernea, Coulter, and Russell 1985; Denning 1985; DenBiggelaar 1991; Röling and Engel 1989; Moris 1991; Fairhead 1990). Working with and through indigenous organizations for development has been addressed by Cook and Grut (1989), Messerschmidt (1991), Rau (1991), Groenfeldt (1991), and Uphoff (1985).

9. The case studies themselves are not discussed here, since they are being compiled in a forthcoming volume edited by the authors. A few of these experiences were presented in the World Bank–sponsored Workshop on Traditional and Modern Approaches to Natural Resource Management in Latin America and the Caribbean, Washington, D.C., 25–26 April 1995.

10. Examples of detailed economic analyses include a long-term case study in western Honduras by Felber and Foletti (1989), which found that green revolution corn technology offered a lower economic return than traditional growing practices. Mausolff and Farber (1995) also compared the economic costs and benefits of chemical and low-purchased-input ecological technologies in two Honduran rural development projects and found that traditional practices based on cover cropping with velvet bean (*Macuna pruriens*) tripled average corn yields from a baseline of roughly 700 kilograms to 2000 kilograms per hectare, using only one-fifth of the commonly applied chemical fertilizer.

11. Wiens (1994) argues that irrigation and water rights are as unequally distributed as landownership in Latin America, and that the effects on efficiency may be as serious as the equity issue. For example, if water rights usurped from the highlands in the Andes were restored and the traditional conservation structures and management institutions were rehabilitated, an estimated 40 percent increase in farm productivity could occur (Wiens 1994).

References

Alexandratos, N., ed. 1988. *World Agriculture Toward 2000: A FAO Study*. London: FAO and Belhaven Press.

Anderson, Simon. 1996. Research Centres and Participatory Research: Issues and Implications (Mexico). Paper presented to the Study Group on Mediating Sustainability, ILAS, London. Revised version.

Ascher, W., and R. Healy. 1990. *Natural Resource Policy-making in Developing Countries: Environment, Economic Growth, and Income Distribution*. Durham, N.C.: Duke University Press.

Ashby, J. 1987. The Effects of Different Types of Farmer Participation on the Management of On-Farm Trials. *Agricultural Administration and Extension* 24: 234–252.

Ashby, J., C. Quirós, T. Gracia, M. Guerrero, and J. Roa. 1990. Farmer Participation Early in the Evaluation of Agricultural Technologies. Paper presented at the seminar Reviving Local Self-Reliance: Challenges for Rural/Regional Development in Eastern and Southern Africa, Arusha, Tanzania, 21–24 February.

Ashby, J., T. Gracia, M. Guerrero, C. Quirós, J. Roa, and J. Beltrán. 1995. Organizing Experimenting Farmers for Participation in Agricultural Research and Technology Development. Paper presented at the Workshop on Traditional and Modern Approaches to Natural Resource Management in Latin America and the Caribbean, World Bank, Washington, D.C., 25–26 April.

Barborak, J., and G. Green. 1987. Implementing the World Conservation Strategy: Success Stories from Central America and Colombia. In *Sustainable Resource Development in the Third World*, edited by D. Southgate and J. Disinger. Boulder, Colo.: Westview Press.

Bebbington, A. 1995. Organising for Change—Organising for Modernization? Campesino Federations and Technical Change in Andean and Amazonian Resource Management. Paper presented at the Workshop on Traditional and Modern Approaches to Natural Resource Management in Latin America and the Caribbean, World Bank, Washington, D.C., 25–26 April.

Bebbington, A., G. Thiele, P. Davies, M. Prager, and H. Riveros, eds. 1993. *Non-Governmental Organizations and the State in Latin America*. London: Routledge.

Bheenick, R. 1989. *Successful Development in Africa: Case Studies of Projects, Programs, and Policies*. EDI Development Policy Case Series, Analytical Case Studies no. 1. Washington, D.C.: World Bank.

Biggs, S., and E. Clay. 1981. Sources of Innovation in Agricultural Technology. *World Development* 9 (4): 321–36.

Bilsborrow, R. 1992. Population Growth, Internal Migration, and Environmental Degradation in Rural Areas of Developing Countries. *European Journal of Population* 8:125–48.

Boserup, E. 1965. *The Conditions of Agricultural Growth*. Chicago: Aldine.

Boserup, E. 1981. *Population and Technology*. Oxford: Blackwell.

Brokensha, D., D. Warren, and O. Werner, eds. 1980. *Indigenous Knowledge Systems and Development*. Lanham, Md.: University Press of America.

Butler, L., and J. Waud. 1990. Strengthening Extension through the Concepts of Farming Systems Research and Extension (FSRE) and Sustainability. *Journal of Farming Systems Research-Extension* 1 (1): 77–92.

Cashman, K. 1989. Agricultural Research Centers and Indigenous Knowledge Systems in a Worldwide Perspective: Where Do We Go from Here? In *Indigenous Knowldge Systems: Implications for Agriculture and International Development*, edited by D. Warren et al. Technology and Social Change Program, Studies in Technology and Social Change no. 11. Ames: Iowa State University.

CEPAL. 1994. *Panorama Social de América Latina*. Comisión Económica para América Latina, Santiago, Chile.

Cernea, M., J. Coulter, and J. Russell. 1985. Building the Research-Extension-Farmer Continuum: Some Current Issues. In *Research-Extension-Farmer: A Two-Way Continuum for Agricultural Development*, edited by M. Cernea, J. Coulter, and J. Russell. Washington, D.C.: World Bank.

CGIAR. 1990. A Possible Expansion of CGIAR. Technical Advisory Committee, Consultative Group on International Agricultural Research, Washington, D.C.

Chambers, R., A. Pacey, and L. Thrupp, eds. 1989. *Farmer First: Farmer Innovation and Agricultural Research*. London: IT Publications.

Chambers, R., and J. Pretty. 1994. Are the International Agricultural Research Centres Tackling the Crucial Issues of Poverty and Sustainability? *International Agricultural Development* (November/December).

Conway, G., and J. Pretty. 1991. *Unwelcome Harvest: Agriculture and Pollution*. London: Earthscan.

Cook, C., and M. Grut. 1989. *Agroforestry in Sub-Saharan Africa: A Farmers' Perspective*. World Bank Technical Paper no. 112. Washington, D.C.: World Bank.

Cortes, J. 1995. Seed Systems Development in Peru. Paper presented at the Workshop on Traditional and Modern Approaches to Natural Resource Management in Latin America and the Caribbean, World Bank, Washington, D.C., 25–26 April.

DeJanvry, A. 1981. *The Agrarian Question and Reformism in Latin America*. Baltimore: Johns Hopkins University Press.

Delgado, C. 1995. *Africa's Changing Agricultural Development Strategies: Past and Present Paradigms as a Guide to the Future*. IFPRI Discussion Paper no. 3. Washington, D.C.: IFPRI.

DenBiggelaar, C. 1991. Farming Systems Development: Synthesizing Indigenous and Scientific Knowledge Systems. *Agriculture and Human Values* 8 (1): 25–36.

Denning, G. 1985. Integrating Agricultural Extension Programs with Farming Systems Research. In *Research-Extension-Farmer: A Two-Way Continuum for Agricultural Development*, edited by M. Cernea, J. Coulter, and J. Russell. Washington, D.C.: World Bank.

DeWalt, B. 1995. Using Indigenous Knowledge to Improve Agriculture and Natural Resource Management. Paper presented at the Workshop on Traditional and Modern Approaches to Natural Resource Management in Latin America and the Caribbean, World Bank, Washington, D.C., 25–26 April.

DeWalt, B., P. Vergne, and M. Hardin. 1993. Population, Aquaculture and Environmental Deterioration: The Gulf of Fonseca, Honduras. Paper prepared for the Rene Dubos Center Forum on Population, Environment, and Development, New York Academy of Medicine, 22–23 September.

Dommen, A. 1988. *Innovation in African Agriculture*. Boulder, Colo.: Westview Press.

Dommen, A. 1989. A Rationale for African Low-Resource Agriculture in Terms of Economic Theory. In *Indigenous Knowledge Systems: Implications for Agriculture and International Development*, edited by D. Warren et al. Technology and Social Change Program, Studies in Technology and Social Change, no. 11. Ames: Iowa State University.

Dover, M., and L. Talbot. 1987. *To Feed the Earth: Agroecology for Sustainable Development*. Washington, D.C.: World Resources Institute.

Fairhead, J. 1990. Fields of Struggle: Towards a Social History of Farming Knowledge and Practice in a Bwisha Community, Kivu, Zaire. Ph.D. diss. SOAS, University of London.

FAO. 1989. *The State of Food and Agriculture*. Rome: FAO.

FAO. 1991. *Sustainable Agriculture and Rural Development in Latin America and the Caribbean*. Regional Document no. 3. Rome: FAO.

FAO. 1992. *FAO Agrotast-PC, Population, Production, and Food Balance Sheets Domains*. Rome: FAO.

FAO. 1993. *Water Policies and Agricultural Development*. Rome: FAO.

Farrington, J., and A. Bebbington. 1994. *From Research to Innovation: Getting the Most from Interaction with NGOs in Farming Systems Research and Extension*. Gatekeeper Series Paper no. 43. London: IIED.

Farrington, J., and A. Martin. 1988. *Farmer Participation in Agricultural Research: A Review of Concepts and Practices*. Agricultural Administration Unit Occasional Paper no. 9. London: ODI.

Fearnside, P. 1986. *Human Carrying Capacity of the Brazilian Rain Forest*. New York: Columbia University Press.

Felber, R., and C. Foletti. 1989. *Estudio sobre Agricultura Migratoria en la Zona de Guajiquiro, Opatoro*. Programa Marcala-Goascoran, Secretaría de Recursos Naturales y Cooperación Suiza del Desarrollo (COSUDE), Marcala, Honduras.

Franke, R., and B. Chasin. 1980. *Seeds of Famine: Ecological Destruction and the Development Dilemma in the West African Sahel*. Montclair, N.J.: Allanheld, Osmun and Co.

Garrett, J. 1995. *A 2020 Vision for Food, Agriculture, and the Environment in Latin America*. IFPRI Discussion Paper no. 6. Washington, D.C.: IFPRI.

Groenfeldt, D. 1991. Building on Tradition: Indigenous Irrigation Knowledge and Sustainable Development in Asia. *Agriculture and Human Values* 8 (1): 114–120.

Hansen, M. 1987. *Escape from the Pesticide Treadmill: Alternative to Pesticides in Developing Countries*. Mount Vernon, N.Y.: Institute for Consumer Policy Research.

Haverkort, B., W. Hiemstra, C. Reijntjes, and S. Essers. 1988. Strengthening Farmers' Capacity for Technology Development. *IIEA Newsletter* 4 (3): 3–7.

Hayami, Y., and V. Ruttan. 1985. *Agricultural Development: An International Perspective*. Baltimore: Johns Hopkins University Press.

Hayami, Y., and V. Ruttan. 1987. Population Growth and Agricultural Productivity. In *Population Growth and Economic Development: Issues and Evidence*, edited by D. G. Johnson and R. E. Lee. Madison: University of Wisconsin Press.

Hazell, P., and C. Ramasamy. 1989. *The Green Revolution Reconsidered: The Impact of the High-Yielding Rice Varieties in South India.* Baltimore: Johns Hopkins University Press.

Hochman, E., and D. Zilberman. 1986. Optimal Strategies of Development Processes of Frontier Environments. *Science of the Total Environment* 55 (1): 111–20.

Hoskins, M. 1984. Observations on Indigenous and Modern Agro-Forestry Activities in West Africa. In *Social, Economic, and Institutional Aspects of Agroforestry,* edited by J. Jackson. Tokyo: United Nations University.

Jaffee, S., and J. Srivastava. 1992. *Seed System Development: The Appropriate Roles of the Private and Public Sectors.* Discussion Paper no. 167. Washington, D.C.: World Bank.

Jodha, N. 1990. Mountain Agriculture: The Search for Sustainability. *Journal of Farming Systems Research-Extension* 1 (1): 55–75.

Kevin, C., and G. Schreiber. 1992. *The Population, Agriculture, Environmental Nexus in Sub-Saharan Africa.* Agriculture and Rural Development Series no. 1. Washington, D.C.: Africa Technical Department, Agriculture Division, World Bank.

Leisinger, K. 1995. *Socio-political Effects of New Biotechnologies in Developing Countries.* IFPRI Discussion Paper no. 2. Washington, D.C.: IFPRI.

Leonard, H., ed. 1989. *Environment and the Poor: Development Strategies for a Common Agenda.* New Brunswick, N.J., and Oxford: Transaction Books.

Marten, G., ed. 1986. *Traditional Agriculture in Southeast Asia: A Human Ecology Perspective.* Boulder, Colo.: Westview Press.

Maurya, D., A. Bottrall, and J. Farrington. 1988. Improved Livelihoods, Genetic Diversity and Farmer Participation: A Strategy for Rice Breeding in Rain-Fed Areas of India. *Experimental Agriculture* 24 (3): 210–19.

Mausolff, C., and S. Farber. 1995. An Economic Analysis of Ecological Agricultural Technologies among Peasant Farmers in Honduras. *Ecological Economics* 12: 237–48.

Mellor, J., and G. Desai, eds. 1985. *Agricultural Change and Rural Poverty.* Baltimore: Johns Hopkins University Press.

Messerschmidt, D. 1991. Community Forestry Management and the Opportunities of Local Traditions: A View from Nepal. In *Indigenous Knowledge Systems: The Cultural Dimension of Development,* edited by D. Warren et al. London: Kegan Paul International.

Moris, J. 1991. *Extension Alternatives in Tropical Africa.* London: ODI.

Niamir, M. 1990. *Community Forestry: Herders' Decision-Making in Natural Resources Management in Arid and Semi-arid Africa.* Community Forest Note no. 4. Rome: FAO.

Norem, R., R. Yoder, and Y. Martin. 1989. Indigenous Agricultural Knowledge and Gender Issues in Third World Agricultural Development. In *Indigenous Knowledge Systems: Implications for Agriculture and International Development,* edited by D. Warren, L. J. Slikkerveer, and S. O. Titilola. Technology and Social Change Program, Studies in Technology and Social Change, no. 11. Ames: Iowa State University.

Oldeman, L., V. van Engelen, and J. Pulles. 1990. The Extent of Human-Induced Soil Degradation. In *World Map of the Status of Human-Induced Soil Degradation: An Explanatory Note,* edited by L. Oldeman, R. Hakkeling, and W. Sombroek. Wageningen, the Netherlands: International Soil Reference and Information Centre.

Pingali, P., and H. Binswanger. 1987. Population Density and Agricultural Intensification: A Study of the Evolution of Technologies in Tropical Agriculture. In

Population Growth and Economic Development: Issues and Evidence, edited by D. G. Johnson and R. E. Lee. Madison: University of Wisconsin Press.

Pinstrup-Andersen, P., and R. Pandya-Lorch. 1995. *Alleviating Poverty, Intensifying Agriculture, and Effectively Managing Natural Resources*. IFPRI Discussion Paper no. 1. Washington, D.C.: IFPRI.

Poffenberger, M., ed. 1990. *Keepers of the Forest: Land Management Alternatives in Southeast Asia*. West Harford, Conn.: Kumarian Press.

Pretty, J. 1994. *Regenerating Agriculture: Policies and Practices for Sustainable Growth and Self-Reliance*. London: Earthscan.

Psacharopoulos, G., and H. Patrinos, eds. 1994. *Indigenous People and Poverty in Latin America: An Empirical Analysis*. World Bank Regional and Sectoral Studies. Washington, D.C.: World Bank.

Rau, B. 1991. *From Feast to Famine: Official Cures and Grassroots Remedies to Africa's Food Crisis*. London: Zed Books.

Reijntjes, C., B. Haverkort, and A. Waters-Bayer. 1992. *Farming for the Future: An Introduction to Low-External-Input and Sustainable Agriculture*. London: Macmillan Press.

Repetto, R. 1985. *Paying the Price: Pesticide Subsidies in Developing Countries*. Research Report no. 2. Washington, D.C.: World Resources Institute.

Rhoades, R. 1984. *Breaking New Ground: Agricultural Anthropology*. Lima: CIP.

Rhoades, R. 1988. Changing Perceptions of Farmers and the Expanding Challenges of International Agricultural Research. Paper presented at the Conference of Farmers and Food Systems, Lima, 26–30 September.

Richards, P. 1985. *Indigenous Agricultural Revolution: Ecology and Food Production in West Africa*. London: Hutchinson.

Rola, A., and P. Pingali. 1993. *Pesticides, Rice Productivity, and Farmers' Health: An Economic Assessment*. Manila and Washington, D.C.: International Rice Research Institute and World Resources Institute.

Röling, N., and P. Engel. 1989. IKS and Knowledge Management: Utilizing Indigenous Knowledge in Institutional Knowledge Systems. In *Indigenous Knowledge Systems: Implications for Agriculture and International Development*, edited by D. Warren, L. J. Slikkerveer, and S. O. Titilola. Technology and Social Change Program, Studies in Technology and Social Change, no. 11. Ames: Iowa State University.

Rosegrant, M., and M. Svendsen. 1993. Asian Food Production in the 1990s: Irrigation Investment and Management Policy. *Food Policy* 18 (February): 13–32.

Scherr, S., and P. Hazell. 1993. Sustainable Agricultural Development Strategies in Fragile Lands. Paper prepared for the American Agricultural Economics Association International Pre-Conference on Post–Green Revolution Agricultural Development Strategies in the Third World: What Next? Orlando, Florida, 30–31 July.

Scoones, I., and J. Thompson, eds. 1994. *Beyond Farmer First: Rural People's Knowledge, Agricultural Research and Extension Practice*. London: IT Publications.

Smith, N. 1995. *Bio-diversity and Agroforestry Along the Amazon Flood-plain*. Paper presented at the Workshop on Traditional and Modern Approaches to Natural Resource Management in Latin America and the Caribbean, World Bank, Washington, D.C., 25–26 April.

Southgate, D. 1992. Promoting Resource Degradation in Latin America: Tropical Deforestation, Soil Erosion, and Coastal Ecosystem Disturbance in Ecuador. *Economic Development and Cultural Change* 40 (4): 787–807.

TAC/CGIAR. 1988. *Sustainable Agricultural Production: Implications for International Agricultural Research*. Rome: FAO.

TAC/CGIAR. 1993. *The Ecoregional Approach to Research in the CGIAR: Report of the TAC/Center Directors Working Group.* Rome: FAO.

Thrupp, L. 1989. Legitimizing Local Knowledge: From Displacement to Empowerment for Third World People. *Agriculture and Human Values* 3:13–25.

Titilola, S. 1990. *The Economics of Incorporating Indigenous Knowledge Systems into Agricultural Development: A Model and Analytical Framework.* Technology and Social Change Program, Studies in Technology and Social Change no. 17. Ames: Iowa State University.

Uphoff, N. 1985. Fitting Projects to People. In *Putting People First: Sociological Variables in Rural Development,* edited by M. Cernea. Oxford and New York: Oxford University Press.

Verhelst, T. 1990. *No Life Without Roots: Culture and Development.* London: Zed Books.

Warren, D. 1991. *Using Indigenous Knowledge in Agricultural Development.* Discussion Paper no. 127. Washington, D.C.: World Bank.

Warren, D. 1995. Using Indigenous Knowledge in Agricultural Development. Paper presented at the Workshop on Traditional and Modern Approaches to Natural Resource Management in Latin America and the Caribbean, World Bank, Washington, D.C., 25–26 April.

Warren, D., and K. Cashman. 1989. *Indigenous Knowledge for Sustainable Agricultural and Rural Development.* Gatekeeper Series Paper no. 10. London: IIED.

Weber, P. 1992. A Place for Pesticides? *Worldwatch* 15 (3): 22–23.

WHO. 1980. *Disease Prevention and Control in Water Development Schemes.* Geneva: WHO.

WHO. 1983. *Environmental Health Impact Assessment of Irrigated Agricultural Development Projects.* Geneva: WHO.

Wiens, T. 1994. *Rural Poverty, Sustainable Natural Resource Management, and Overall Rural Development in the LAC Region: The World Bank's Strategy.* Latin America and the Caribbean Technical Department, Background Note. Washington, D.C.: World Bank.

Wiggins, S., and E. Cromwell. 1995. NGOs and Seed Provision to Smallholders in Developing Countries. *World Development* 23 (3): 413–22.

Wilcox, B. 1991. In Situ Conservation of Genetic Resources: Institutional and Technical Requirements for a Global System and the Need for Strengthening Capacity at the National Level. Report prepared for the Forestry Division of the Food and Agriculture Organization of the United Nations, Rome.

Wilcox, B. 1995. Rural Development and Indigenous Resources: Toward a Geographic-Based Assessment Framework. Paper presented at the Workshop on Traditional and Modern Approaches to Natural Resource Management in Latin America and the Caribbean, World Bank, Washington, D.C., 25–26 April.

Wilken, G. 1987. *Good Farmers: Traditional Agricultural Resource Management in Mexico and Central America.* Berkeley: University of California Press.

Wolf, E. 1986. *Beyond the Green Revolution: New Approaches for Third World Agriculture.* Washington, D.C.: Worldwatch Institute.

World Bank 1992. *World Development Report 1992.* New York: Oxford University Press.

World Bank. 1993. *Towards a Bank Strategy for Agriculture in the Latin America and Caribbean Region.* Washington, D.C.: Latin America and the Caribbean Technical Department, World Bank.

2

NGOs

Mediators of Sustainability/ Intermediaries in Transition?

Anthony Bebbington

Two profound and related changes are reworking the rural economy of the Andean region today and must necessarily be central considerations in the search for sustainable forms of development in the region. One is an institutional change: the reform of the state, the increased assertiveness of civil society, and the ever-increasing space being given to and expected of the private sector. The other is the progressive removal of subsidies, tariffs, quotas, and trade barriers that together are intended to create a more favorable environment for investment and private-sector activity. These latter changes are linked closely to the institutional changes, in that they represent an attempt to increase the role of the marketplace in mediating patterns of development and at the same time reduce the relative importance of the role of government in this process of mediation.

Within this context, more is being asked of nongovernmental organizations (NGOs).[1] They are expected to assume some of the roles previously played by the state, as well as roles typically performed by commercial organizations—all in the name of a more sustainable, participatory, and efficient development (see World Bank 1995; Farrington and Bebbington 1993). Yet this new context itself, and the sea change in development thinking of which it is indicative, presents important—indeed, penetrating—challenges to Latin American NGOs that work in rural development. Furthermore, these challenges are being presented at a time when NGOs are faced with a series of institutional problems—problems that are characterized in this chapter as crises of legitimacy, identity, and sustainability.

Recent discussions have begun to raise some of these issues at a general level (Edwards and Hulme 1996; Hulme and Edwards 1997). This chapter takes the discussion to a more specific level, that of the Andes and Chile. The argument is that changes in the political economy of Andean America have demanded that NGOs rethink their relationships with the state and the market. In turn, this rethinking has triggered a series of more general uncertainties about the role of NGOs in development. These uncertainties are only part of a larger crisis in alternative development thinking and a related uncertainty about the legitimate (and most effective) roles of civil society, the state, and the market in development. However, this chapter also argues that these uncertainties—in the context of a funding crisis—are fostering a set of institutional changes among NGOs that, though painful, offer the possibility of rerooting civil society institutions back into their own societies, such that they are better adapted to the conditions of their own societies and less distorted by the incentives and agendas fostered by foreign aid. This, in turn, has implications for how we think of civil society's role in development and, more practically, for how donors might best support this process of institutional adjustment.

The chapter first outlines challenges faced by Andean NGOs in this current context and the conditions of emerging crisis in which NGOs find themselves. The chapter then outlines forms of institutional response that appear to be emerging among NGOs as they seek to identify new roles for themselves in rural development and new, sustainable institutional forms that allow them to continue to exist as NGOs.[2]

Contexts of Cha[lle]nge in the Rural Sector

An Emerging Exclusionary Development?

Although macroeconomic indicators in the Andean region suggest a situation of relative economic stability and growth, other indicators suggest that certain groups are not benefiting from this (Sotomayor 1995). The implication is that a significant part of Andean America is following an exclusionary form of development in which little new employment is generated and in which neither private investors nor governments demonstrate much interest in the popular economy as a cornerstone of the development process (Figueroa, Altamirano, and Sulmont 1996).

Related to this process is the ongoing transformation of the rural economy, with legislation allowing the free sale of formerly community-controlled resources, and a market liberalization that leads traditional domestic products to be displaced by imports from both Latin America and further afield. In some areas, levels of rural out-migration continue to grow, and even if tight urban labor markets lead some of these migrants to return to the countryside, a massive return seems unlikely.[3]

The viability of the contemporary small farm (campesino) economy is thus in question unless new forms of generating, capturing, and reinvesting wealth in rural areas are found. A recent study of the Peruvian and Bolivian Andes concludes: "If the market is the determining factor in the definition of rural policy, Andean agriculture has two possibilities: to disappear, or to modernize violently in order to achieve competitive levels of productivity and production" (van Niekerk 1994, 319).

In this context, the impact of state and NGO development programs on campesino livelihood and economy has generally been minimal. Notwithstanding evidence of positive social and institutional impact, van Niekerk (1994) estimates that for every dollar spent in NGO projects in the Andes, the return has been sixty cents (that is, negative). In Chile, a recent econometric study of the national technical assistance program suggests that the program has had no impact on rural livelihoods (López 1995). Much remains to be done to define rural interventions that might enhance the economic foundations of rural sustainability. In the final instance, if NGOs do not achieve such an economic impact, they will have failed in their mission (Torranzo 1995).

Government Reforms and Changing NGO-State Relationships

The redefinition of the role of government in Latin America has significantly changed the relationships between NGOs and the state. The new Latin American state is shifting away from direct implementation of development initiatives; increasingly, it subcontracts or finances programs implemented by nonstate institutions and plays a normative role: setting and monitoring the rules of the game and creating an environment in which private enterprise and civic initiative can flourish.

This shift is double edged. On the one hand, it creates a basis from which a more efficient and potentially more accountable state can be built, and upon which the scope for civil society and NGO initiatives can be expanded (Sotomayor 1994). Indeed, for years, NGOs have criticized the state for its bureaucracy, inefficiency, and exclusion of civil society organizations from decision-making processes (see Clark 1991). On the other hand, one senses that this redefinition of the state's role is also part of a more profound redefinition of the social contract between the state and elite and the popular sectors (Bebbington and Thiele 1993; Edwards and Hulme 1996). Furthermore, these public-sector reforms go hand in hand with economic changes that may weaken civil society organizations and thus undermine their potential to take advantage of these new spaces for participation (Pearce 1993). What do these changes imply for NGOs?

A New Franchise State?

Geoff Wood (1997) has described the emergence of a franchise state for the case of Bangladesh—a state that subcontracts the delivery of services, the management of activities and resources, and so forth. Although the terminology may be extreme for Latin America, it captures the sense of a fundamental shift in the role of the state. In some cases, for all intents and purposes, the state has withdrawn from both the financing and the delivery of services: examples would be rural credit in Peru and agricultural extension in highland Bolivia. In other cases, the state does not implement activities but, through the use of contracts and grants, "puts out" a range of delivery and management activities to the private sector—commercial and nongovernmental. In Bolivia, NGOs are now sought out to manage national parks, reserves, and protected areas. In Chile—since the mid-1980s—agricultural extension has been subcontracted to the private sector, and NGOs and farmers' organizations have been allowed to bid for these contracts since the end of the Pinochet regime; in 1993–94 NGOs were on contract to attend some 35 percent of the farm families served by the Institute of Agricultural and Livestock Development (INDAP) (Berdegué 1994; INDAP 1995). Other Latin American countries are initiating variants of this bidding and subcontracting model for the delivery of rural development services, and they look to NGOs—as do their financing agencies—to assume a significant share of these contracts. In many cases, the state (and donors) assume that these NGOs will take on this role and do so under the coordination of the state.

Funds, Funds, and More Funds

One of the more significant institutional innovations associated with adjustment in Latin America has been the emergence of special funds. The first of these was the Social Emergency Fund in Bolivia from 1986 to 1989 (Wurgaft 1992). This experience inspired a plethora of other social funds throughout Latin America (and Africa), as well as other types of funds for the environment, campesino development, regional development, and so forth.[4] These funds may be independent of line ministries, but not always of politics, and they serve to channel resources through intermediary organizations to implement development activities. The government—and often donors—sets norms and rules for the funds and may also influence the details of how activities funded this way function. This is perhaps most significant in the case of credit operations funded by so-called second-level funds inside government and channeled through NGOs. In these cases, the government influences the interest rate that the NGO charges and may place the NGO under the supervision of the regulatory agency for the banking sector.[5]

These new funds are important for NGOs. At a time of declining external financial support, the funds offer new domestic sources of finance—but sources that come with conditions that may imply new ways of operating for the NGO (for instance, market interest rates) and more intimate forms of state supervision of NGO operations than has previously been the case.

Municipalization, Decentralization, and Local-Level Governance

The role of local government in development has been growing steadily, albeit to varying extents across the region. Legislation for fiscal and administrative decentralization has passed unprecedented development resources to local governments in countries such as Bolivia and Colombia—even though in many cases the new responsibilities of local governments go beyond the new resource bases placed at their disposal. As municipalities, for the most part, lack any implementation capacity, these resources are used to subcontract private and nongovernmental institutions to implement programs. In Bolivia, this decentralization legislation—in the guise of the Participation Law (Ley de Participación Popular)—also begins to require NGOs to coordinate their development activities with municipal development plans (Péres 1994). Either way, this increased role of local government requires NGOs to deal with local government at largely unprecedented levels. This ought to be a positive development for NGOs, because in general, their experience of coordination with the state has been more positive at the local level than at the central and ministerial levels. However, this change may also require NGOs to rethink and to decentralize their own institutional structures. Moreover, where local elites dominate local government, this decentralization can create a more hostile environment for local civil society actors (see Fox and Aranda 1996).

The Challenge of Democratization

The progressive transition to democracy in Latin America has had significant influence on the relation between NGOs and the state. Democratization initially led to optimism among NGOs (especially among those linked to the political parties that were elected into power) that they were going to be able to participate in the definition of policy. In practice, the spaces for such participation have been more restricted than had been hoped. In some cases, governments have continued to be quite distrusting of NGOs; in other cases, such as in Chile and Colombia, where state capacity is strong, they have not accorded NGOs as much importance as had been hoped. Indeed, in some cases, governments now question the legitimacy of NGOs to participate in such discussions.

The transition to elected government has also left traditional NGO critiques of the state somewhat outdated. Indeed, financing agencies that were previously sympathetic to this posture of critique and rejection may now be less inclined to support NGOs with this attitude. In the end, democratization demands that NGOs reconsider their own governance structures (see Tandon 1995). Now it is the NGO, rather than government, that is the unelected (indeed, self-elected) institution. Democracy has not been the positive experience that NGOs had hoped it might be.

New States, New Donors

Not only have changes in the nature of the state challenged NGOs to rethink their roles and relationships, the practices of donor agencies have also forced this change. Many have encouraged increased engagement between NGOs and government. In some cases, such as Chile, donors that previously financed NGOs have shifted their support to government, channeling funds to the types of financing mechanism through which governments subcontract or finance proposals of NGOs and other private actors. At the same time, it is now standard practice for bilateral and multilateral supported projects to include components in which project activities will be implemented by NGOs. These shifts in the channeling and conditions of development financing have been a direct incentive (indeed, a force) to encourage NGOs to deal more closely with government.

And yet, as experiences accumulate, there is also a growing critical attitude toward NGOs that is evident both in documents and in corridor discussions: that they should modernize, professionalize, and democratize themselves (see Edwards and Hulme 1996). This may ultimately lead to a fall in official funding for NGOs, with what remains being channeled to them via government. According to the project coordinator for the International Fund for Agricultural Development (IFAD) in Bolivia, Peru, and Argentina:

> There has been a reduction in the pressure on IFAD to increase the extent to which their projects are implemented by NGOs. In part this is because donor governments have ever diminishing resources dedicated to rural development, and perhaps also because projects implemented by NGOs have not proved to be significantly more efficient than those implemented by governments (Haudry de Soucy 1994, 4).

The State of Play

NGOs' experiences in these new relationships with government have been mixed. The pressures on NGOs to deal more closely with the state seem difficult to resist. Yet they do not appear to offer NGOs a new niche as key actors in the discussion and formulation of policy—the tendency has been to exclude them from these arenas. NGOs have found more scope in the role of joint NGO-state implementation and coordination of development. Yet this type of relationship presents NGOs with many conundrums regarding their role and identity. To the extent that governments and donors tend to instrumentalize NGOs within policy and program frameworks developed without NGO involvement, this dilemma is especially acute.

Complicating the situation even more is the sense that NGO-state relationships tend to be more effective and more intimate when NGO and government share party political affinities and the sense that the availability of new development funds channeled through the state for subcontracting has encouraged the emergence of a gamut of organizations that call themselves NGOs but are to all intents and purposes commercial ventures or survival strategies for professionals. These opportunist NGOs undermine the legitimacy of the label "NGO."

Crisis Tendencies Among Latin American NGOs

The combined effect of these changes has been to catalyze and aggravate the crises of identity, legitimacy, and sustainability that NGOs face in this new policy context. These crises threaten the ability of these organizations to play any future role in Latin American social transformation toward sustainable rural development.

The NGO Identity Crisis

The crisis of identity that afflicts many of Latin America's NGOs[6] operating in the rural arena has different hues but common origins. It is an effect of the change in the dominant political-economic model in Latin America, of the NGOs' loss of their own model for sustainable and progressive social change, and of a certain cultural change within NGOs themselves.

Neo-liberalism and formal democracy have been around for a number of years now. It seems, however, that NGOs still have been unable to forge for themselves a new role and a new identity within this changed policy context. The elements of this change are well known. In the 1970s and early 1980s, the dominant model in much of Latin America was of a bureaucratic, authoritarian, often repressive, socially exclusive, but quite interventionist state. Under this context, NGOs forged themselves a coherent identity. They were the *organizaciones revindicativas*, organizations that resisted the state and existed to support the popular sectors under the model and to propose alternative forms of development. To the extent that they generated alternative models, these models presupposed a strong and interventionist state, but under a different form of social control—one that was more participatory. With the changes discussed in the preceding section, however, NGOs can no longer base an identity in resistance and demands, and their former state-centered alternatives are no longer relevant.

Complicating this situation further for NGOs is that in this new context, a number of new and apparently interesting opportunities have opened up for NGOs to participate in program design and implementation. Initially, NGOs were enthusiastic about this—no more so than in Chile in 1990 and

1991. But then came the disappointment. Many soon concluded that these spaces for dialogue were limited. Governments incorporated a number of NGO ideas (for example, in the Law of Popular Participation in Bolivia and in the agricultural programs of Chile and Colombia), and then the scope for policy dialogue closed.

In the process, NGOs have found themselves becoming implementers of other people's programs—primarily of the state and bi- and multilateral donors. This, they realize, cannot be the basis of an institutional identity—it makes them no more than consultants. Although many see the necessity for some form of relationship with the state and donors (Bebbington and Thiele 1993; idc, various years), the difficulty is in knowing how to manage these relationships to avoid becoming instruments of other institutions' agendas.

Arguably, this uncertainty regarding an appropriate relationship with the state derives from a general NGO uncertainty about development models and the appropriate roles of market, state, and civil society in a future Latin America. In this uncertainty, they are no different from other institutions. But as actors whose notions of development were central to their institutional identities, this uncertainty leads to a lack of clarity about NGO mission, role, and identity.

Linked to this uncertainty about models is a debate within NGOs about how far institutional modernization and professionalization would corrupt their identities.[7] Forces for this modernization are real within many NGOs and come primarily from the younger NGOs and professionals, who insist on the need for more analysis of impact, for more monitoring and evaluation, for more technical expertise, and so forth. Older hands (who also tend to control management positions in these NGOs) argue that such changes have to be treated with care—that in modernizing and professionalizing, NGOs may lose their utopian visions and their capacity to question and propose. Just as indigenous peoples' organizations are challenged to redefine their identities under modernization (Bebbington 1996), so too are NGOs.

The NGO Legitimacy Crisis

Related to this NGO identity crisis is a profound crisis of legitimacy. Popular organizations at all levels, from communities to national campesino confederations, criticize NGOs. To the extent that NGOs traditionally drew on their relationships with the popular sectors for their legitimacy, these critiques are highly significant. Some of these criticisms are of the culture and structure of NGOs—namely, that NGOs refuse to be transparent and to let go of control of projects and resources; that they exclude popular organizations from positions of power and forums of policy dialogue; that they have no right to claim to represent the popular sectors; and that, in the end, they have the same social origins as those who have always dominated the poor. Other criticisms are more operational: that only a small proportion of their funds reach the field; that they earn too much, and that NGOs are technically weak.

At the same time, society at large, which traditionally knew little about NGOs, now understands them much more. And what it understands, it has begun to criticize. In newspaper articles and editorials and in special publications, one increasingly encounters criticisms made in the public domain that NGOs earn large wages from international funds, are unaccountable to society, and engage in subversive activities. One measure of the degree of this social criticism is the barrage of criticism recently dealt out by the Bolivian press. Business also complains that NGOs are not efficient and enjoy special protection through their tax privileges, which makes it difficult for business to compete with these organizations for certain subcontracts (Méndez 1994; IPE 1995). NGOs' duty-free imports also encourage NGOs not to buy locally.

Finally, having been on the defensive for several years, the Latin American state is also increasingly vocal in its criticisms of NGOs. Bolivia again is one of the most recent examples of this, particularly in the debate surrounding the popular participation legislation (Balcazar 1994). In a publication summarizing these discussions, one can find the criticisms that NGOs are not accountable, and that this is unacceptable under a democratic and modernizing state; that NGO activity is often uncoordinated chaos; and that NGOs are inefficient. The implication is that NGO activity should be coordinated and supervised by the state.

So NGO legitimacy is being questioned on all sides. Indeed, it seems that NGOs find themselves in the unfortunate position that their principal base of legitimacy is their donors (and even here, there are more and more criticisms). This only enhances the sense that NGOs are institutional intrusions rather than institutions that are grounded in their own societies.

The NGO Sustainability Crisis

This lack of a social base in their own societies means that NGOs are not sustainable social actors. Their action depends on finances from other countries and, increasingly, from their own states. The tendency is for external funds to diminish—in particular, the more flexible forms of finance (de la Maza 1995). In this situation, to survive, NGOs will have to be increasingly dependent on governments and multilateral donors. This implies deepening the very relationships that many argue undermine the identity of NGOs—relationships that turn NGOs more and more into implementers of other people's programs and that pull them further away from the popular sectors.

Pathways Out of Crisis?

Not only is the Latin American NGO world in some form of crisis, it is also in transition. This transition is being forced by shifts in the structure of donor financing, by the related looming financial problems faced by NGOs, by the changing role of the state in rural development, by NGOs' sensed need to renew their models and strategies in the face of the limited impact they have had, and by a broader cultural change in the development sector. The transition is less a consequence of any sense that NGOs need to reestablish the bases of their legitimacy vis-à-vis the popular sectors and their missions, even though this is arguably the most critical challenge NGOs need to respond to, because it questions their very raison d'être.

Of course, different NGOs have responded and will respond to these pressures for change in different ways, and in different countries the legislation may favor particular forms of institutional constitution to a greater or lesser extent. The resulting adjustments are in part institutional changes that one can already perceive and in part predictions of the types of changes that we might expect to see. The pathways outlined below are not intended to be mutually exclusive—a number of NGOs combine, in differing degrees, these different institutional adjustments.

From NGO to Consulting Group

As the pressure on NGOs' traditional sources of finance grows, it seems likely that they will look increasingly toward the contracts and special funds that have become available from government and donor agencies and that are in many cases an effect of the progressive redefinition of the state's role in the provision of development services. This process will inevitably lead to NGOs having closer relationships with their governments.

This shift toward implementing programs designed to a large extent to reflect the objectives of government and donor agencies implies a shift in the nature of the NGO, turning it—at least within the realm of these contracts—into a subcontracted development consultancy. Particularly in the case of those NGOs contracted to implement field activities, the difference between them and commercial subcontractees becomes blurred, and maintenance of the label NGO, and the related tax-exempt status, will become more difficult to justify.

Other NGOs that begin to depend on contracted research and advisory work will have more flexibility and autonomy than straight development consultancies. However, they will not have the flexibility of independently and core financed research groups, nor the sorts of links to popular sectors that Latin American NGOs historically possessed.

For those NGOs that move along this trajectory, then, their relationship with the state will become one of subcontractee, adviser, implementer. Although some may be able to question dominant institutions to a certain degree, their scope for this may be limited by their financial dependence on these same sources. This would parallel the experience of U.K. Save the Children, which has found that its space to criticize the British government has become more constricted as it has received more government funding for its activities (Bell 1996).[8]

Whatever the drawbacks of this option, it addresses the crises outlined earlier. It begins to resolve the problem of identity—the organization becomes a private consulting group working within, and with resources deriving from, the dominant policy framework. It also resolves the problem of legitimacy, in that the organization lives or dies on the basis of the quality of its work—the contract becomes the mechanism through which the NGO becomes accountable (to the contractor). Finally, problems of financial sustainability begin to be addressed through the income from contracts, although many NGOs would have to increase their efficiency, staff quality, and other factors in order to survive in this way.

From NGO to Social Enterprise

It has been argued strenuously that one of the most legitimate and appropriate options facing NGOs is to become a form of social enterprise (Zadek and Gatwood 1995). One option is for the NGO to engage in market operations in order to generate a profit that can be used for the NGO's development work. Another option is for the NGO to do its development by engaging with the market; in this option, the NGO works with poor people with the common goal of improving their access to the market. Ultimately, the objective may be to change the nature of the market through the example of being a different sort of market actor (Chapter 6 of this book; Zadek 1995; Zadek and Gatwood 1995). Each of these paths toward social enterprise (and other paths in between) are apparent among a limited number of NGOs in the region.

As an example of the first path, we can look to Chile: when Chilean NGOs began bidding for contracts from the national agricultural development institute, INDAP, for the provision of technical assistance to small farmers, some hoped that this contracted form of service provision would be sufficiently profitable to subsidize some of the NGOs' other activities. In practice, this seems rarely to have been possible—the income from the contracts merely covers the operating costs of implementing the contracts, and the NGO has sometimes had to subsidize the contracted activity. Other examples of this strategy are NGOs that run print shops, publication businesses, even funeral parlors to generate income for their social development operations (Bebbington and Rivera 1994).

This first option is primarily concerned with building the basis for institutional sustainability; it is not necessarily an element in a deeper rethinking of the bases for sustainable development. The social enterprise path can also be part of such a theoretical reformulation, however. Slowly and unevenly, NGOs are coming to the conclusion that any option for sustainable rural development from the popular economy must be built on increased production of wealth and on a renegotiation of market relationships so that the wealth produced is more likely to be reinvested within the region in which it was produced. The need to renegotiate relationships underlying the markets for the inputs and the products of the popular economy, coupled with a search for ways of generating new incomes for NGOs, has also led some organizations along the path to social enterprise. In these cases, the creation of the social enterprise is seen not only as a means of institutional sustainability but also as a mechanism that is necessary to change markets and stimulate more inclusive development in the areas in which the enterprise operates.

Some of these enterprise-like NGOs have existed for a long time, especially in product marketing—for example, Commercializando Como Hermanos (MCCH) and the Tiendas Camiri in Ecuador, and Candela and Antisuyo in Peru. It has been more difficult to build social enterprise models around the marketing of inputs, such as the provision of technical assistance services to poor farmers. One exception, though, is the provision of rural financial services—particularly credit. As a result of recent changes in banking laws, a growing number of NGOs are moving into this sector, individually or in consortia. The attraction of moving into this sector is that financial services (credit, savings, insurance, and so forth) are certainly sought by many rural people, and to the extent that NGOs are seeking new instruments and approaches to enhance their impact on rural poverty and livelihoods, financial service provision at a significant scale has offered one potential option. More pragmatically, these are services that people are willing to pay for, and under certain program designs, the rates of repayment have been very high.

In Bolivia, for instance, the last decade has seen the emergence of a number of large NGO-based credit operations. Examples of operations involving consortia of NGOs include the National Ecumenical Development Association (ANED), a group of some twenty-five NGOs and churches that began to work in financial services in the late 1980s, and the Alternative Development Foundation (FADES) created by seven NGOs in 1988. In other cases, individual NGOs have developed substantial credit programs, such as the Fund for Communal Development (FONDECO) created by CIPCA (but now independent), and Sartawi, which began its credit program in 1991 (Rojas 1995). Over the years, these different credit programs have grown significantly—

**Table 2.1   Credit Portfolios of NGOs Providing Financial Services
in Bolivia (Late 1994)**

Institution	Accumulated Lending (US$1,000)	Current Loans (US$1,000)
ANED	3,553	1,700
FADES	4,758	2,185
FONDECO	4,352	3,111
Sartawi	3,500	1,100

Source: Adapted from Rojas 1995.

partly in an attempt to seek financial sustainability by spreading overheads over larger portfolios and thus reducing operating costs (Table 2.1).

Rojas (1995) estimates that NGOs in Bolivia have lent some $32 million as credit since they began operations, and that against the 165 rural branches that private commercial banks have, NGOs and related institutions already have some 45 branches—some 20 percent of all rural credit-giving branch offices. The tendency seems to be that these credit-providing NGOs will continue to grow as opportunities to access funds from central banking institutions increase. The quid pro quo, however, will probably be that as availability of funds from second-tier banks increases, so will the extent to which the state supervises these credit-giving NGOs via the Superintendencia de Bancos.

In some sense, this path offers a response to the dimensions of NGO crisis outlined earlier. The identity of the NGO becomes that of an institution ensuring that financial and economic services reach the rural poor on a massive scale—even if in doing so the NGO is assuming a role that is in some sense functional to the broader model of structural adjustment. The legitimacy of the NGO comes from its efficacy and efficiency in delivering such services, and its institutional sustainability comes through the income generated from performing these financial services. Again, though, this type of institution is no longer an NGO in the historic sense of the term. It is an institution that is neither state nor market, that combines a commercial and a social logic in its operation, and that is thus able to play a role, sustainably, that neither the state nor the market can play.

This response brings with it clear challenges. One is the administrative and economic efficiency required to be financially sustainable not simply as an enterprise but as a *social* enterprise whose profit margins are even lower. Another challenge is that this approach—particularly that of the large-scale credit NGO—carries the seeds of a model of development that is quite different from the models that NGOs have traditionally espoused.[9] It thus implies a different sense of institutional mission for the NGOs that follow this path.

From NGO to Popular Organization

In the rush to find new roles for NGOs, in the midst of the language of reengineering, strategic plans, and institutional modernization, others worry that important babies have been thrown out along with the muddy bath water. It may be that demanding rights (*revindicar*), being isolated from the state, and being excessively ideological are now inappropriate stances. But it may also be that there were other qualities of being an NGO in those early years of resistance to authoritarianism and repression that embody the essence of being an NGO today. Many argue that this essence lies in playing the role of innovator—innovating from a strong basis in the popular sectors, and in a way that strengthens popular organizations.

This return to the roots is another response evident among a number of NGOs. In this case, the emphasis is less on financial survival and more on identifying and restoring the founding principles that are still relevant to the 1990s. This response to the crisis grows out of the notion that both the legitimacy and the identity of NGOs are grounded in their relationship with the popular sectors.[10] It is also implied that the NGO is not a legitimate actor in its own right, with its own agenda, but rather that its role is to support popular organizations in the elaboration of development alternatives that the popular sectors would carry forward (and not the NGO).

Of course, recovering their legitimacy vis-à-vis the popular sectors is not enough. NGOs also have to recover this legitimacy in relation to state and society at large, so that these proposals have greater legitimacy and so that the NGO may play some role in facilitating more creative relationships between state, popular organizations and society. This implies that in this path, too, the NGO has to be more professional (so that these alternatives are feasible and convincing) and transparent.

This path, however, leaves two great questions hanging: how will the challenge of financial sustainability be addressed, and where might these alternative proposals be concentrated?

Financing a Return to the Roots

In the short term, it may be that this third pathway could be financed by the same donors that have traditionally sponsored more radical approaches to development. In the longer term, however, alternative financing strategies are necessary: strategies that are more organically linked to NGOs themselves and to the societies of which they are a part. Only in this way will the pursuit of alternatives be buffered from its dependence on international relationships, fads, and fashions.

This is easier said than done. Funds channeled via the public sector are unlikely to sponsor work that in many regards will be critical of dominant policy. One alternative is once again that of the social enterprise—a mechanism for the self-financing of this strategy. In this sense, there is scope to combine the three pathways: the NGO would have legal existence as a business and as an NGO, and the business activities would cross finance development activities. A number of NGOs have begun to take this route, though so far, it allows only partial financing of their nonprofit activities.

Another potential strategy is to create civil society financing mechanisms within Latin American countries. There is some precedent for this. One precedent, of course, is from the North, where foundations (such as the Ford Foundation) play this role. Within Latin America, there are similar, though fewer, mechanisms through which businesses and other sources have created endowed funds. Of course, for this to occur on any large scale will require a change in the business culture in the region. Latin American business is not famed for its social responsibility—although recent initiatives suggest that some change is afoot.[11] There are also precedents in which donor agencies have contributed to the endowment of special funds for the environment, showing that even though it is not easy for donors to contribute to endowments (as opposed to year-to-year projects), it is possible.[12] These endowments offer a model for rethinking civil society financing too. Ultimately, only if donors can move from project funding to building and endowing domestic financing institutions will their resources have any sustainable impact in the region.

What Alternatives?

Some years ago, Marc Nerfin (1987) laid out four principles of alternative development: focusing on needs orientation, fostering self-reliance, promoting ecological sustainability, and empowering people to transform their societies. Although attractive in sentiment, these principles do not seem grounded or pragmatic enough to respond to the current challenges facing the campesino and popular economies. As others note, these social alternatives ultimately depend on a material base (Peet and Watts 1996). Yet very little attention has been paid to thinking through alternative forms of growth from the popular economy that would offer such a material base—an alternative to the current trend toward exclusionary growth. It is therefore in the areas of production, inclusive growth, and rural income and employment generation that alternatives need to be sought. Indeed, if we have learned anything from thirty years of rural development, it is the importance of identifying mechanisms for the capture and productive reinvestment in rural areas of the value deriving from rural economic activities (de Janvry and Sadoulet 1988; Klein 1993).

The alternatives that most urgently demand NGO attention are somewhat different from the alternatives of the past. Rather than rethinking politics, the challenge is to experiment in new relationships between the campesino economy and private enterprise. Over the past decade, NGOs have had to rethink their relationship with the state. In the next decade, they must rethink their notions of and relationships toward the market and private capital.

The current challenges encountered by the NGO sector in South America draw our attention to the links between political economy and the composition of actors in civil society. The rise of the NGO sector in South America was driven in large measure by the emergence of particular political and economic regimes, and by an international context in which Northern states and civil societies were able and willing to channel resources to this non-governmental institutional response. The current crises in the NGO sector are likewise being driven in large measure by political and economic shifts in South America—adjustment, democratization, and public-sector reform—and in Europe.

These crises are painful, but they may lead to the recomposition of an NGO sector that is more sustainable and whose different institutional reforms reflect more appropriate responses to the wider institutional environment of which they are a part. Although it would be going too far to argue that the NGO sector is a creature and a creation of Northern financing agencies, it is true that NGOs are dependent on external resources and, to some extent—except in their earliest moments—have come to represent institutional forms that have been induced by external relationships rather than by relationships within their own societies. In that sense, in their present form they are institutions that cannot possibly be sustainable.

It is in the context of this situation that we should understand both the current crises of and the transitions within the NGO sector. In the absence of a (much-needed) development of domestic, endowed, autonomous civil society funding mechanisms in South America, the move toward becoming consultant groups, social enterprises or financial service institutions represents financing options through which these organizations can move themselves onto more sustainable ground. This is so in part because these institutional forms bridge gaps between campesino livelihoods, state, and market that would otherwise remain because of the current logic of state and market actors. While dealing with the fundamental problem of financial sustainability, these new forms also help resolve the problems of legitimacy and identity that would otherwise threaten the coherence and viability of these organizations. Of course, such institutional shifts will require profound transformations within the NGO sector—changes of legal status, efficiency, professionalism, attitude, and identity.

The third pathway—returning to the roots—will also require shifts. It will require NGOs to go back to their original mission and establish closer relationships with popular organizations: relationships of accountability, trust, responsiveness, and solidarity. In this sense, this route—like the other possible paths—also involves addressing certain institutional "distortions" in the sector—distortions in which NGOs have become more distant from those popular sectors than they were at their inception. This institutional adjustment will similarly help address the problems of identity and legitimacy, though in very different ways. It will be far harder for it to resolve problems of financial sustainability, largely because the popular sectors are not able to finance such NGOs. This is the type of NGO that will continue to need the support of traditional NGO donors, and it is perhaps the type in which future donor support could best be concentrated.

NGOs have always been in part a response to state failure, in part a response to market failure, and in part a response to weaknesses in popular organizations. On some occasions, they acted more like a state; at other times, more like an ally of the popular organization; and on other (rarer) occasions, more like a (socially oriented) market actor. This mixing of roles was never easy and has contributed to the crises that NGOs now face. In the future, one senses that NGOs will increasingly have to focus on only one of these primary roles, and this will clarify the bases of their identity, legitimacy, and financial security. As they do this, many organizations following the first or second pathway will cease to be NGOs in the historical sense of the term and should cease to be called such. This should not necessarily be seen as a problem. Indeed, it will probably be the way in which the validity and legitimacy of the denomination NGO will be sustained.

Notes

An earlier version of this chapter was published in Spanish as "Crisis y Caminos: Reflexiones heréticas acerca de las ONG, el estado y un desarrollo rural sustentable en América Latina," published by the NGO program NOGUB, of the Swiss Development Corporation in Bolivia. An earlier version in English was "New States: New NGOs?" published in *World Development* 25, no. 11 (November 1997): 1755–65, copyright 1997, and we are grateful for permission from Elsevier Science to reproduce it here. Of the many people whose thoughts on this theme have influenced mine, I would like to thank Thomas F. Carroll, Teresa Domingo, Denise Humphreys, Adalberto Kopp, Walter Milligan, Diego Muñoz, Nico van Niekerk, Jose Antonio Péres, Galo Ramón, Roger Riddell, Dagoberto Rivera, and Octavio Sotomayor.

   1. The definition of NGOs always demands some clarification. In this chapter, I am using the term to refer to private, professionally staffed, nonmembership and intermediary development organizations—what Carroll (1992) calls grassroots support organizations (GSOs).

2. This chapter is the outcome not of a single research project but of a collection of separate pieces of research, evaluation, and advisory work primarily in the Andean region and Chile, but also in Central America (see Bebbington, Carrasco, et al. 1992; Bebbington and Thiele 1993; Bebbington and Rivera 1994; Bebbington et al. 1995; and idc, various years).

3. The exception would be in areas where out-migration was as much an effect of violence as of economic hardships, and where large numbers of migrants may return once levels of violence subside.

4. Not only has the notion of national funds taken hold, but projects now increasingly incorporate special funds for subcontracting or financing activities of civil society organizations.

5. This was the case, for example, in the creation of EDPYMEs (Entidades de Desarrollo para la Pequeña y Mediana Empresa) in Peru or the FFPs (Fondos Financieros Privados) in Bolivia.

6. See note 1.

7. A particularly acute instance of this debate has been the recent discussions of reengineering in the NGO DESCO, in Peru.

8. The U.S. Congress's recent suggestion that nonprofits that receive state funding should not be allowed to engage in any sort of advocacy is another Northern parallel.

9. For instance, in order to be sustainable, self-financing credit funds tend to work with larger loans and with campesinos who are not the poorest. They also tend to return to the same lenders each year as a means of reducing transaction costs (Schmidt and Zeitinger 1995).

10. The consultancy and social enterprise responses seek neither NGO legitimacy nor identity in this relationship with the popular sectors.

11. Initiatives include the creation of the Grupo 2021 in Peru; the social responsibility work of PDVSA, the national petroleum company of Venezuela; the 1995 international conference on business and social responsibility in Colombia; and publications of the World Business Council for Sustainable Development.

12. For instance, the trust fund created in Bolivia for financing the recurrent costs of the National System of Protected Areas.

References

Balcazar, R. M. 1994. Introducción. In *La Participación Popular y las ONG*. La Paz, Bolivia: Dirección de Coordinación con las ONG, Ministerio de Desarrollo Humano.

Bebbington, A. 1996. Movements, Modernizations and Markets: Indigenous Organizations and Agrarian Strategies in Ecuador. In *Liberation Ecologies: Environment, Development and Social Movements*, edited by R. Peet and M. Watts. London and New York: Routledge.

Bebbington, A., H. Carrasco, L. Peralvo, G. Ramón, V. Torres, and J. Trujillo. 1992. *Los Actores de Una Decada Ganada: Tribus, Comunas y Campesinos en la Modernidad*. Quito, Ecuador: Comunidec/Abya-Yala.

Bebbington, A., T. Domingo, A. Kopp, and J. Quisbert. 1995. *Campesino Federations, Food Systems and Rural Politics in Bolivia*. London: IIED.

Bebbington, A., and D. Rivera. 1994. *Nicaragua: Finnida NGO Support Program Evaluation Country Case Study*. Helsinki: Ministry for Foreign Affairs.

Bebbington, A., and G. Thiele. 1993. *NGOs and the State in Latin America: Rethinking Roles in Sustainable Agricultural Development*. London and New York: Routledge.

Bell, W. 1996. The Relationship Between NGO Service Provision and Advocacy: The UK Experience. Manuscript.

Berdegué, J. 1994. El sistema privatizado de extensión agrícola en Chile: 17 años de experiencia. Presented at the International Symposium for Farming Systems Research and Rural Development, Montpellier, France, 21–25 November.

Carroll, T. 1992. *Intermediary NGOs: The Supporting Link in Grassroots Development*. West Hartford, Conn.: Kumarian Press.

Clark, J. 1991. *Democratizing Development: The Role of Voluntary Organizations*. West Hartford, Conn.: Kumarian Press.

de Janvry, A., and E. Sadoulet. 1988. *Investment Strategies to Combat Rural Poverty: A Proposal for Latin America*. Mimeo. Berkeley: Department of Agricultural and Resource Economics, University of California.

de la Maza, G. 1995. Las ONG Chilenas y la Nueva Cooperación. *idc* 2:6.

Edwards, M., and D. Hulme. 1996. Too Close for Comfort: NGOs, the State and Donors. *World Development* 26 (6): 961–73.

Farrington, J., and A. Bebbington. 1993. Reluctant Partners? NGOs, the State and Sustainable Agricultural Development. London and New York: Routledge.

Figueroa, A., T. Altamirano, and D. Sulmont. 1996. *Social Exclusion and Inequality in Peru*. Geneva: International Labor Organization.

Fox, J., and J. Aranda. 1996. *Decentralization and Rural Development in Mexico: Community Participation in Oaxaca's Municipal Funds Program*. Monograph Series 42. San Diego: Center for U.S.-Mexico Studies, University of California.

Haudry de Soucy, R. 1994. ONG, estado y recursos para campesinos desde las instituciones financieras internacionales. *idc* 1:4. (Quotation translated by A. Bebbington.)

Hulme, D., and M. Edwards, eds. 1997. *Too Close for Comfort? NGOs, Donors and the State*. Oxford: Macmillan.

idc. Various years. *Instituciones, Desarrollo, Cooperación: Taller de Intercambio Sobre Relaciones Inter-institucionales en el Sector Rural de América Latina*. Santiago, Chile: GIA.

INDAP. 1995. *Plan de Modernización. Programa de Transferencia Tecnológica (PTT): 1995–1997*. Santiago, Chile: Instituto de Desarrollo Agropecuario.

IPE. 1995. Eficiencia de las ONG's: Solo interés de los financiadores? *Información Política y Económica* 31 (3 July).

Klein, E. 1993. El empleo rural no-agrícola en América Latina. In *Latinoamérica agraria hacia el siglo XXI*. Quito, Ecuador: CEPLAES.

López, R. 1995. *Determinants of Rural Poverty: A Quantitative Analysis for Chile*. Washington, D.C.: Technical Department, Rural Poverty and Natural Resources, Latin America Region, World Bank.

Méndez, A. 1994. Análisis sobre las Organizaciones No-Gubernamentales de Desarrollo (ONG) y su Proyecto de Ley. Mimeo. Tarija: CEPB.

Nerfin, M. 1987. Neither Prince Nor Merchant: Citizen—An Introduction to the Third System. *Development Dialogue* 1987 (1): 170–95.

Pearce, J. 1993. NGOs and Social Change: Agents or Facilitators? *Development in Practice* 3 (3): 222–27.

Peet, R., and M. Watts. 1996. Liberation Ecology: Development, Sustainability, and the Environment in an Age of Market Triumphalism. In *Liberation Ecologies:*

74 ■ Policies and Actors in Mediation

Environment, Development, Social Movements, edited by R. Peet and M. Watts. London and New York: Routledge.

Péres, J. A. 1994. Bolivia: Cambios en Políticas y Oportunidades para las Principales Corrientes de Enfoques Participativos en el Desarrollo Rural Sustentable. Paper presented to the Study Group on Mediating Sustainability, IIED, London, meeting in Wageningen, Netherlands, November.

Rojas, E. R. 1995. Crédito Rural: Una Aproximación a la Oferta. La Paz, Bolivia: Centro de Estudios y Proyectos.

Schmidt, R. H., and C. P. Zeitinger. 1995. Prospects, Problems and Potential of Credit Rating NGOs. Mimeo. Frankfurt am Main: IPC.

Sotomayor, O. 1994. Chile: la necesidad de consolidar vinculos ONG-gobierno. idc 1:2.

Sotomayor, O. 1995. Políticas campesinas de fomento productivo y condicionantes macroeconómicas en Chile: dilemas para la decada del 90. Manuscript. Santiago, Chile: Grupo de Investigaciones Agrarias.

Tandon, R. 1996. Board Games: Governance and Accountability in NGOs. In Beyond the Magic Bullet: NGO Performance and Accountability in the Post–Cold War World, edited by M. Edwards and D. Hulme. West Hartford, Conn.: Kumarian Press.

Torranzo, C. 1995. Los Retos para Las ONG. idc 2:6.

van Niekerk, N. 1994. Desarrollo Rural en los Andes. Un Estudio sobre los Programas de Desarrollo de Organizaciones no Gubernamentales. Leiden Development Studies no. 13. Leiden: Leiden University.

Wood, G. 1997. States Without Citizens: The Problem of the Franchise State. In Too Close for Comfort? NGOs, Donors and the State, edited by D. Hulme and M. Edwards. Oxford: Macmillan.

World Bank. 1995. How the World Bank Works with NGOs. Washington, D.C.: World Bank.

Wurgaft, J. 1992. Social Investment Funds and Economic Restructuring in Latin America. International Labor Review 131 (1): 35–44.

Zadek, S. 1995. Integrated Approaches to Sustainability. Principles and Principals. Paper presented to the Study Group on Mediating Sustainability, Institute of Latin American Studies, London, June.

Zadek, S., and M. Gatwood. 1995. Transforming the Transnational NGOs: Social Auditing or Bust? In Beyond the Magic Bullet: NGO Performance and Accountability in the Post–Cold War World, edited by M. Edwards and D. Hulme. West Hartford, Conn.: Kumarian Press.

Part II

Measurement as Mediation

3

Widening the Lens on Impact Assessment

The Inter-American Foundation and Its Grassroots Development Framework—The Cone

Marion Ritchey-Vance

In the early 1990s, the Inter-American Foundation (IAF), like most funding agencies, was challenged to present the results of its support of grassroots development. At the time, the foundation was entering its third decade and struggling to articulate what difference twenty years of funding had made in the lives of its beneficiaries. It was clear in the field that the whole was far greater than the sum of ad hoc small-scale projects. The results included expanding the frontiers of civil society and translating micro programs into policy change at the community, regional, or even national level. What was not clear yet to foundation and local organization staff was how to pinpoint, document, and communicate the civic and policy dimensions of hundreds of community initiatives.

IAF was originally conceived by the U.S. Congress in the late 1960s as an experimental alternative to massive government-to-government development programs that tended to transfer technological models for implementation in the field. The enabling legislation set broad parameters that were fleshed out by a visionary board of directors and turned into specific action by the early officers and staff. In brief, the foundation set out to be a responsive agency aiming to entertain grant proposals from nongovernmental organizations (NGOs) on their own terms; the NGOs would then carry out programs that they themselves had designed.

The foundation was not only to fund projects but also to encourage and facilitate documentation and dissemination of the lessons learned. Over the years, IAF explored various approaches to evaluation, consciously trying to weigh process as well as product. None quite supplanted the traditional system of reporting on results, which remained geared to the conventional sectoral measures of jobs created, houses built, acres tilled. More substantive issues were reserved for case histories, on an anecdotal, project-by-project basis.

In 1988, IAF experimented with a methodology designed to meet its dual needs—on the one hand, to tell the story of its grantees in a style that captured the complexity of the change process at the grassroots, and on the

Table 3.1   IAF Beneficiary Assessment, 1988–90: Sociopolitical
Impact Indicators

Area of Inquiry	Category
Performance of the grantee organization	■ Beneficiaries (who, how many, circumstances) ■ Accountability and participation (decision making, management, implementation of activities) ■ Beneficiary/client satisfaction (timeliness, reliability, transparency) ■ Contagion effect (organizationally, technology, methodology—replicability) ■ Adaptation to change ■ Organizational sustainability (financial solidity and managerial capacity) ■ Future plans (pointing to group cohesiveness)
Impact of the grant	■ Impact on individuals and families (income, production, access to credit, tenure security, reduced need to migrate, self-dignity, confidence) ■ Organizational impacts (control over markets, awareness of impact of self-help efforts, organizational capacity, institutional legitimacy)
Cost effectiveness	■ Organizational costs vs. economic returns (internal rate of return, net present value, increased production, employment, income and access to credit, time savings)

Source: Adapted from Humphreys 1994, 4–5.

other hand, to respond to growing pressures from government oversight
agencies such as the Office of Management and Budget. An outside evaluation team contracted by the foundation set out to collect hard data as well as
qualitative material on the impact of IAF-funded projects. The team applied
a set of sociopolitical performance indicators (Table 3.1), based on beneficiary assessment methods used widely in the late 1980s (Avina et al. 1990;
Humphreys 1994).

As assessed by one of the evaluators (Humphreys 1994, 3), the methodology was useful in illuminating issues related to internal dynamics and leadership and in considering organizational activities and evolution from a
historical perspective. It established measurable indicators and a potential
means for self-analysis of an organization's impact. However, the team recognized serious weaknesses in this process. Because there was no systematic
cross-checking with beneficiaries or against original mandates, grantees
could obscure weak performance by simply stating success. Time and cost
intensiveness reduced the viability of this method for all organizations
involved. And, finally, this evaluation method did not allow the organizations'

and beneficiaries' experience to be set within a regional or national context or program strategy. IAF also realized that relations among different actors needed to be given more consideration and that the three key areas of inquiry (cost effectiveness, impact of grant, performance of grantee) needed to be appraised more systematically.

The next major initiative occurred in the early 1990s, when a newly arrived management team at IAF called for an inventory of results of the foundation's funding. The question of what to document and how came to a head. Experience had clearly shown that results of a given project ran the gamut from sacks of beans produced or number of children innoculated to the creation of community-based organizations capable of marshaling and managing funds or the increased capacity of those organizations to influence policy decisions. The inventory of results as conceived and announced, how-ever, was tilted heavily toward the material, tangible, and convertible-to-cash achievements. Veteran staff began casting about for a way to focus attention on the full array of results—institutional, civic, and political as well as mate-rial. They also hoped to find a methodology that would allow the appraisal to underscore the importance of the intangible gains that often determine whether a development process will be sustainable over time. The composi-tional impact was to be considered—the interaction of parts within the whole—rather than simply the accumulatory impact—the sum of a larger number of parts. Thus was born the Cone, later baptized with the more for-mal title of Grassroots Development Framework (Zaffaroni 1997).

Concept

Initially, the purpose of the Cone was to broaden conceptual horizons within the foundation itself—to widen the lens through which the organization looked for results. The search was for a tool that would focus on organiza-tional and communitywide impact as well as material products, and that would facilitate systematic assessment of grantees' work in a wider context.

The basic structure of the resulting framework is simple (Figure 3.1). It essentially codifies IAF's practical experience that grassroots development produces results on three levels—direct benefits to families and individuals, strengthening of organizations, and broader impact in the community or society at large—and that these results are of two kinds, tangible and intan-gible. Tangible in this context refers to the material results that can be counted, measured, and substantiated by direct evidence. Intangible results are those gains or losses that can be observed and inferred but are harder to measure directly or in simple quantitative ways.

The graphic does not pretend to be a cross section of society. The coni-cal shape is meant simply to represent the widening impact of grassroots

Figure 3.1 The Cone: Basic Structure

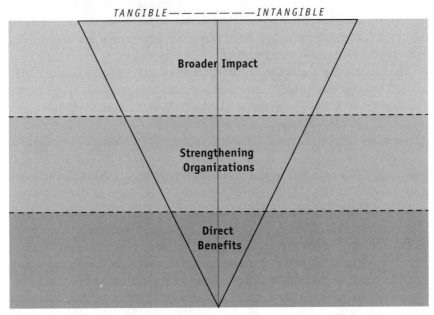

Source: Inter-American Foundation.

initiatives upward from the individual through the organization to the community level. At the first level, benefits to individual families are the most closely associated with a given project but are limited in scope. At the second and third levels, the effects—in increased capacity, sustained effort, and numbers of people reached over time—can be considerably greater, but the project is only one of a number of contributing factors.

The basic structure of the Cone, with its three levels and two types of impact, forms six windows, or categories, which encompass the principal kinds of results or effects that have been noted by foundation representatives and documented in grantee reports and monitors' evaluations over IAF's twenty-year experience (Figure 3.2). By way of cross-checking the validity of the categories, IAF consulted with field staff and grantees. In the fall of 1993, the categories were tested specifically against a major evaluation already under way in Costa Rica. The purpose was to see whether the results of that evaluation—conducted in an open-ended, iterative fashion—coincided with the categories as defined by the Cone.

Based on the Costa Rican experience and general consultation, consensus on the basic structure of the Cone was arrived at rather readily. In general, practitioners identified with the schematic format and felt that it reflected reality on the ground. Much more difficult to come by was agreement on

Figure 3.2 The Cone: Six Categories of Impact

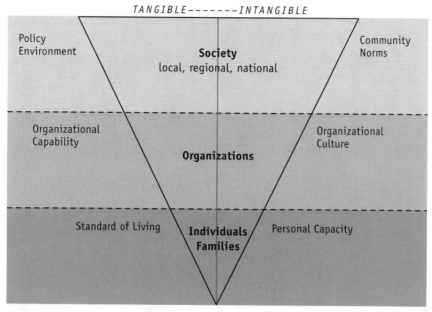

Source: Inter-American Foundation.

how the categories should be entitled for easy reference. The currently existing titles were taken from suggestions by representatives and in-country support teams and have been accepted as working titles but are subject to revision once they have been tried (Figure 3.3.).

Variables

Each of the categories is defined by a set of concepts or variables. The logic of the Cone is to progress from clearly defined concepts (categories, then variables) to the level of specific indicators. *Only when the concepts and definitions were clear and there was broad consensus about what was important to measure did IAF shift attention to specific indicators of progress or accomplishment in each of the variables.*

Variables are intended to focus on outcomes, not activities (such as number of training courses held). The twenty-two variables ultimately selected for the trial phase of implementation in the field are the product of months of debate among IAF staff, its partner organizations in Latin America and the Caribbean, and the in-country support (ICS) teams.

In effect, it was an exercise in defining the essence—trying to distill from the ether of development theory and practice the substance that seasoned practitioners considered key. Although the idea was to expand the field of

Figure 3.3 The Logic of the Cone: Categories and Variables

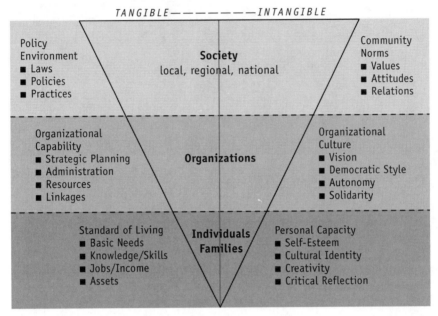

TANGIBLE———————INTANGIBLE

Policy
Environment
■ Laws
■ Policies
■ Practices

Society
local, regional, national

Community
Norms
■ Values
■ Attitudes
■ Relations

Organizational
Capability
■ Strategic Planning
■ Administration
■ Resources
■ Linkages

Organizations

Organizational
Culture
■ Vision
■ Democratic Style
■ Autonomy
■ Solidarity

Standard of Living
■ Basic Needs
■ Knowledge/Skills
■ Jobs/Income
■ Assets

**Individuals
Families**

Personal Capacity
■ Self-Esteem
■ Cultural Identity
■ Creativity
■ Critical Reflection

Source: Inter-American Foundation.

inquiry, IAF evaluation staff were clear that only a limited number of variables could be tracked in each category, lest the whole enterprise collapse of its own weight.

The exercise of whittling down and honing the set of variables and building a degree of consensus through endless meetings, exchanges of correspondence, and revisions consumed the lion's share of the start-up phase. It was also the critical step in the process. The by-product of the laborious effort to unify criteria is a nascent common language (mutually agreed-upon meanings for given terms) that has greatly facilitated communication between IAF and its extended family of ICS teams and grantees. As an ICS director remarked after a major evaluation seminar in Costa Rica: "[t]hough most of us had never met, we were able to enter almost immediately into a substantive discussion on shared terms."

Based on feedback and experience gleaned from the eighteen-month trial phase in the field (1995–96), the variables will be modified and redefined as necessary.

Indicators

The next step was to begin to identify and define indicators. Work on indicators began in 1994 and was in a preliminary test phase until the end of

1996. In the process of field-testing the Cone, the four pilot ICS groups had drawn up their respective sets of indicators. In one geographic region (Southern Cone), the four country teams came together to produce a single, unified version. That process was carried a step further in February 1995, when IAF invited key personnel from each of the four pilot ICS teams (from Ecuador, Uruguay, the Dominican Republic, and Costa Rica) to Washington for an intensive three-day meeting with IAF staff to hammer out a first, experimental global menu of indicators.

To keep the system from becoming unwieldy under the weight of some 1,000 active projects, the decision was made early on that the central database at IAF would record a maximum of five indicators per variable, including two that are standard for all projects and up to three additional at the discretion of the ICS team or the grantee. The standard indicators are of necessity general in nature, to accommodate the wide variety of projects. The discretionary indicators can be tailored in each case to the characteristics of a particular project. For each indicator, the menu specifies description, type, unit of analysis, and unit of measure and requests specific examples or illustrations to substantiate the information. Beyond the two standard and three discretionary indicators that will be entered into ICS and IAF-wide databases, any ICS team is free to create and record additional indicators as time and budget allow.

During 1996, the testing of indicators was in its preliminary phase. Based on feedback gathered from staff, ICS teams, grantees, and beneficiaries, the indicators (and, if necessary, the variables) will be revised and a second version of the menu of indicators produced.

The Mega-variables

Participation, empowerment, and sustainability are basic concepts or tenets of the foundation. Although they form the conceptual undergirding of the framework, they do not appear explicitly. The framework intentionally disaggregates these abstract concepts into more concrete and measurable components. Achieving sustainability, for example, involves building personal and organizational capacities; it requires the ability to mobilize and administer resources, and it implies changes in prevailing policies and attitudes and space to function in the system. By laying out openly the complexity of such mega-variables as sustainability or empowerment, the framework helps illustrate how a single-minded focus on short-term, quantifiable, bottom-line results can in fact undermine the long-term goal of sustainable development.

A major premise of the framework is that all its facets are interconnected and that there is constant interaction among them. Although graphic representation hardens the lines, the gradation between *tangible* and *intangible* and between *individual* and *organization* or *organization* and *community* is in fact a continuum rather than a hard-and-fast distinction.

Balance between the variables on the tangible and intangible sides is key. Strategic planning capability, for example, is an important achievement at the organizational level. But if it is not accompanied by a clear vision of where the organization is going, planning becomes a sterile exercise. Mobilization of resources is key, but if the organization compromises its autonomy in the process, it may become ineffective in the long run.

What's Behind the Framework?

As staff and ICS teams began to work with the Cone, its conceptual underpinnings have been debated and fleshed out. A common query in the early stages was *¿Qué hay detrás de esto?* What's behind the framework?

Numerous assumptions and hypotheses underlie the framework. They reflect IAF's founding values as well as what it has learned in practice from its grantees and colleagues in Latin America and the Caribbean over twenty-five years. Some of those assumptions are:

■ Local action is essential, but it is not sufficient.

■ Sustainable development requires *change* in institutions and in the rules of the game—laws, policies, practices, attitudes.

■ NGOs, grassroots organizations, and networks can be vital links between people and policy, but only so long as the NGO is the means and does not become an end in itself.

■ The energy that drives grassroots development springs from the complex interplay among material, social, and cultural aspirations.

■ The process is *not* linear; it is not a simple matter of going from Problem A to Solution B.

Application

From the Drawing Board to the Field

Once the basic concepts of the Grassroots Development Framework had been debated and largely accepted in-house, IAF sponsored pilot tests in the field. Four of the foundation's ICS teams—in Uruguay, Ecuador, the Dominican Republic, and Costa Rica—took on the task of turning an idea into a fledgling system. In the process, they became the principal architects of the indicators, the methodologies, and the instruments that took the framework from the drawing board to the field.

Soon after the pilot phase was initiated, more than a dozen other ICS teams, notably in Colombia, Honduras, Chile, Panama, El Salvador, Brazil,

Venezuela, Bolivia, Nicaragua, and Mexico, began experimenting with various facets of the framework—incorporating it into planning and monitoring and taking it to the next step, to the grantees.

Response from Grantees.   Clearly, not every grantee is enchanted with the idea of a new IAF-proposed format, but the following observation from an ICS director in the Southern Cone is echoed in other regions as well: "What has most struck me [in the process of applying the framework] is the degree of interest and assimilation on the part of grantees . . . in general, they see it as a real contribution. . . . The response exceeded our most optimistic expectations."

The degree of interest *is* a striking departure from the weary resignation that tends to greet new demands for information. One reason is identification with the holistic nature of the framework. "This is what we've been wanting from funders for a long time," remarked a Dominican NGO leader. "It's the first time the intangible things are taken into account. That's what most of the effort goes into, but we've never known how to record the results." A longtime grantee in Cartagena said simply, "It gives us a way to see the worth of our work."

This degree of acceptance by grantees of the Cone methodology has its roots less in the method or concept than in the approach taken by the ICS teams in their work with grantees. The emphasis of ICS collaboration in the trials of this evaluation method has been on grantee involvement, not donor enforcement: making sure that the Cone is not perceived as just another *peaje institucional* or toll that grantees have to pay for being in a donor's portfolio. As one ICS team made clear in the Andean regional workshop:

> There are two ways of applying the Cone. One is to get the boxes checked and the forms filled in, and the other is make sure that grantees have the opportunity to really appropriate this tool so that it becomes a genuine part of their own planning and evaluation process.

In fact, different grantees are finding different uses for the Cone. Noting that analysis and communication of results had always proved complicated in their line of work, a Southern Cone grantee observed:

> The Cone helps us organize ideas and define what information is relevant and to summarize data and observations clearly and precisely. . . . Systematic and participatory analysis of results helped us correct course and increase our impact . . . we found ourselves rethinking the design of the project.

Perhaps the most encouraging sign that the framework can be more than *peaje institucional* is its spillover in several countries to projects beyond the IAF orbit. Foro Juvenil, a respected grantee group in Uruguay, has worked actively with the ICS team to integrate the framework as a basis for identifying and analyzing the results of its job training program. "Once we really master the Cone, as a framework and as a tool, it will give us the means to set up and manage our own information system . . . tailored if need be to our own particular vision." After a year's trial, Foro Juvenil staff opted to extend the framework to three other major programs, though none is funded by IAF. Why? A senior staff member argued: "The framework suits us because of what we are: a project to transfer methodologies and influence policy. We need an instrument that asks something other than how many youths do you serve?"

Pilot Application

One of the earliest pilot tests was carried out by the Ecuador ICS, COMU-NIDEC. "Cheap, simple, agile, participatory" was how the veteran learning coordinator summed up the framework after a year's total immersion. He now adds "adaptable" as a result of recent contract work to tailor the framework to different subject matter, such as the prevention of AIDS and the evaluation of the national Social Investment Fund.

The twenty-four organizations included in the Ecuador study had received foundation funding over a period of eight years or more. In all, they cover a universe of 803,650 direct or potential beneficiaries who represent 7.6 percent of the population of the country—11 percent of those defined as poor. The sample, chosen to be representative of the ethnic and geographic distribution of the Ecuadorean portfolio, includes fourteen grassroots organizations, nine support institutions, and one quasi-governmental agency.

These organizations, in turn, fit into the backdrop of a burgeoning civil society. Like many Latin American countries, Ecuador shows rising macroeconomic indicators but simultaneous worsening of the quality of life for the majority of its citizens. Despite or because of the deteriorating situation, writes the Ecuador ICS team, "new actors are emerging . . . tradesmen, microentrepreneurs, the indigenous, women, youth, and ecologists have formed some 20,000 associations involving 4 million people." They are organized to solve practical problems, to protect natural resources and the environment, and to influence policy decisions that affect their lives.

The backdrop gives significance to statistics from the study that show that 80 percent of the beneficiaries contributed voluntary time or labor to a project, 72 percent acquired new knowledge of civic rights and responsibilities, and 74 percent were linked into some larger network. Nearly 80 percent cited (positive) change in behaviors and attitudes in their environs, including greater tolerance for ethnic diversity, easing of racial and religious tensions, and better coordination among development agencies.

Results of a Pilot Application

The results registered in Ecuador are distilled from a variety of sources, including existing files and reports, key observers and third parties, individual interviews, and workshops. The principal instrument was the *taller participativo*, or group discussion technique, pioneered by COMUNIDEC to elicit input from all segments of the community and encourage reflection and analysis as a basis for ongoing self-evaluation. In preparation, the ICS team conducting the study in Ecuador "translated" the framework into a set of questions (*preguntas generadoras*) to stimulate discussion. They then entered the results into a database and devised a set of parameters that enabled them to analyze the data by category, variable, indicator, and development approach to discern patterns and trends.

Grassroots Groups. Of the fourteen grassroots groups in the Ecuador study, eleven are Indian federations. Like indigenous organizations throughout the hemisphere, Ecuadorean groups are recovering the symbols, customs, and cultural adhesives of civilizations that thrived before the conquest. Although the grassroots organizations included in the study are far from homogeneous, UOCACI—the Union of *Campesino* Organizations of Cicalpa—is illustrative of the organizational makeup and of the kinds of results obtained by the Indian federations, which represent nearly half the sample universe.

UOCACI is composed of thirty-seven community-based organizations in the province of Chimborazo, one of the chronically impoverished regions in Ecuador. Member organizations range from agricultural cooperatives, artisan centers, and mothers' clubs to traditional music groups. In all, they represent 1,432 families—some 7,000 persons. The federation's role includes advocacy and negotiation with the provincial government for services such as water, roads, and electricity, as well as stimulation of local production and development through training, credit, and technical assistance.

Direct Benefits (Level 1). Technical assistance and credit offered by UOCACI reached 70 percent of its membership. The combination of credit, better understanding of the market, and access to a network of community stores resulted in an increase in income for half the beneficiaries. Installation of electricity yielded an estimated savings of 87 percent in fuel costs. At least one-third of the members are applying new agricultural techniques and administrative skills and have a solid grounding in civic rights and responsibilities. In addition to creating thirty-seven jobs, UOCACI mobilized 5,441 person-days of volunteer labor.

Some of the most striking effects are in personal capacities. Self-esteem is not just a nice by-product of development in Chimborazo. It is fundamental to it. Stripped of ancestral land, language, and leadership, the indigenous have lived a feudal life and ridden in the back of the bus for centuries. The word *indio* still carries overtones of humiliation. Two decades of basic education through the federation that recognizes and builds on their native Quichua language, the revitalization of cultural identity and traditions, and concrete gains in services and income have restored the foundations of respect. Self-esteem is reflected in a dramatic drop in alcohol consumption, also encouraged by the evangelical church. The majority of the members express their opinions, and at least a third of them propose solutions to problems and take action.

Organization (Level 2). UOCACI got high marks for its democratic work style and accountability to members. It articulates a clear vision of its role, which is widely shared. The majority of decisions are made by consensus, and there is freedom to express dissenting views.

Twenty members have been trained as managers, and 120 (many of them at the community level) have acquired administrative skills. The result is an integrated leadership that represents both local and regional interests. The organization holds regular assemblies and planning sessions, and information is readily available to any member. UOCACI is linked to a dense network of local, provincial, and national organizations, which gives it access to decision-making bodies.

Although UOCACI has leveraged substantial local funding for projects (for example, 2 billion sucres for the installation of electricity), leaders recognize that the organization still depends heavily on international funders for its operating expenses.

Broader Impact (Level 3). Beyond the projects themselves, UOCACI has played an important role in the larger community—regulating water use, extending the road system, and helping to fashion the new national agrarian law.

One of UOCACI's main contributions is helping to mediate a long-standing conflict between the evangelical church and the ethnic movements in the province and promoting working relationships among formerly hostile organizations. It has also kindled interest among its membership in protecting the environment, particularly by curbing the loss of topsoil.

UOCACI cites significant progress in ethnic and race relations in general and greater tolerance for diversity, but it has a considerable distance to go in improving relationships with government entities.

NGOs. Fundación Natura, one of the nine NGOs or support institutions covered in the Ecuador study, serves as a good illustration of the role of NGOs in fostering civic responsibility and safeguarding things (such as air) that belong neither to the state nor to private individuals but are vital to everyone's well-being.

Fundación Natura was founded by Ecuadorean professionals concerned with the deterioration of the nation's natural resources. Its main thrust is to raise public awareness of deforestation and erosion. Natura operates with a small salaried staff supported by an active board of directors and a fleet of volunteers. Funding comes from its local membership of 800, from the Ecuadorean private sector, and from international agencies.

Direct Benefits (Level 1).   Fundación Natura has reached 50 percent of its potential audience of 20,000 with its environmental awareness campaign. Technical assistance has resulted in controlling the use of pesticides and diagnosing and treating cases of toxic poisoning. In addition to specific agricultural techniques, 50 percent of the beneficiaries learned management and dissemination skills. The search for nontoxic farming methods has sparked a revival of traditional campesino lore and stimulated creativity among small farmers. Thirty-four percent of the beneficiary population is using alternative methods of farming and measuring yields.

Organization (Level 2).   Fundación Natura's well-trained staff conducts strategic planning on a regular basis. The organization's leadership style—with emphasis on teamwork, shared decision making, and respect for minority points of view—sets an example for the organizations it works with. Natura maintains close ties with numerous other Ecuadorean organizations and is a member of ten worldwide networks. Although a large portion of its funding still comes from international donors, Natura has been more successful than most in establishing a local membership base and raising funds from the private sector.

Broader Impact (Level 3).   Through national and international forums and a series of publications, Fundación Natura has succeeded in drawing the nation's attention to the problems of pesticides, deforestation, and erosion. It drafted and helped pass laws regulating the production, importation, marketing, and use of pesticides and other chemical products. These will have a significant long-term impact nationwide.

By marshaling citizens, businesspeople, and policymakers to cooperate on environmental problems that affect them all, Natura has helped bridge cultural barriers and break down stereotypes.

Patterns

Data from the Ecuador study are still preliminary, but some interesting patterns have emerged. The results documented are distributed relatively evenly among the six categories of the framework and are divided almost equally between tangible and intangible (47 and 53 percent, respectively). ICS staff found that the framework helps visualize the various strategies employed by grantees. The NGOs appear to concentrate on organizational strengthening as a means for delivering benefits at the grassroots or for influencing policy and changing norms. The grassroots organizations begin with the basic

building block of personal capacities as a means to encourage a more democratic culture and ultimately affect values and attitudes. Other types of
grantees target policy change directly and work for measures that will be felt
on a regional or national scale.

Foundationwide, an initial survey of results based on a limited number of
variables produced some interesting statistics.

- *Mobilization of resources.* Over the years, the foundation has reported its
  leverage capacity at $1.50 generated for each dollar invested. Data collected
  with the framework show an average return of $2.50 to the dollar, with fig-
  ures as high as $4.30 in the Southern Cone.

- *Practices.* Combined figures for the four geographic regions show that
  methodologies pioneered by IAF grantees have been replicated or adapted
  by 6,024 NGOs and by 1,186 public-sector organizations. The best practices reported were in work with youth and women and in the fields of preventive health, environment, and training. Collectively, IAF grantees have
  shared methodologies with international organizations in some 200
  instances.

- *Relations.* Finally, and of particular interest in this era of democratization
  and decentralization, is the degree to which the NGOs supported by the
  foundation are interacting with and influencing governments, from the
  municipal level to national ministries. Over three-quarters of the organizations show an increase in capacity to negotiate with both the public and the
  private sectors.

Beauty and the (Potential) Beast

To one degree or another, the Grassroots Development Framework is now
being applied in all twenty countries where the foundation works. Results
data are entered and stored in ICS and IAF databases through specialized
software designed by the foundation. Birth pains notwithstanding, the
framework *has* captured the imagination of many staff, ICS teams, and
grantees who understand that civics and policy influence count and are seeking
a way to document it.

Attractions

Attractions of the framework are:

- *Conceptual clarity.*
- *Simplicity.* As described by the Brazil ICS team, "it is a 'simple' concept,
  that allows NGOs and grassroots groups—and even the benefited people—
  to adopt it as part of their institutional life."
- *Visual, graphic presentation.*

- *Flexibility.* It can be adjusted to the context and adapts to different methodologies of use.

- *Vitality.* This springs from broad participation and a three-year dialogue among staff and ICS teams that built the system.

- *Versatility.* It can be applied in broad brush strokes to an entire portfolio or in detail to a given project.

Caveats and Contraindications

Used in the spirit in which it was intended, the framework can be an invigorating tonic. Like all remedies, however, it can cause harmful side effects if not administered carefully. The warning signs are:

- Use as a prescriptive device rather than an organizing principle.

- Straitjacket or Procrustean bed syndrome, in which reality is made to fit into rigid boxes.

- Rote compliance, or box-checking, rather than seeking to broaden understanding.

- Overcomplication, in the form of a proliferation of variables and indicators.

- Excess zeal in aggregating numbers—parodied effectively by a U.S. cooperative that hung the following sign on its door:

Dogwood Crafters

Founded	1974
Members	67
Altitude	2300
Total	4341

- The tail wags the dog—skewing projects and programs to accommodate "the system."

Feedback from Field Tests

The eighteen-month pilot phase alternated field tests with workshops and feedback sessions, which gave the ICS teams the benefit of one another's experience and flagged the major gaps and ambiguities in the system for IAF staff.

The unexpected richness of project profiles that emerged from applying the framework dispelled much of the initial skepticism. "We discovered," said the Colombian learning coordinator, "that the Cone is a prism. Before, we were seeing only the white light. Now we see a full spectrum of color." A similar response came from the ICS team in Brazil: "In some cases, it helped visualize facets of the project of which we were completely unaware." The Brazil team was among the first to use the framework as a tool for "marketing" grassroots experience and influencing policy.

The field tests also pointed out a number of concerns and operational problems. Virtually everyone underscored the importance of *context* as a backdrop and a basis for interpreting results, and of understanding the *strategy*, or *process* employed. In applying the framework, a clear distinction needs to be made between NGOs, or intermediary organizations, and grassroots groups. The *time frame* also continues to be a major question. In a process that often spans a decade or more, how does one deal with results that become manifest well after a grant has ended?

More work was clearly needed (and is currently under way) to develop adequate indicators for productive enterprises and to ensure that issues of gender are taken into account. Creative proxies must be developed to gauge fulfillment of basic needs. There was general agreement that trying to attribute results to a given grant is not practical or germane. More relevant is an understanding of the relationship among results, strategy, and context, which will help get a handle on which development approaches work under what circumstances.

Because some of the measures necessarily rely on informed opinion, there are questions of objectivity and reliability of data. Surprisingly, several ICS teams report that data become more reliable as grantees participate in framing indicators and take the framework to heart to help them improve their own performance. Nonetheless, reliability remains a major concern for ICS teams, which compensate for the opinion factor by consulting a variety of sources, including outsiders and knowledgeable third parties; cross-checking data over time; and contracting full-fledged evaluations where warranted. In addition, the foundation will conduct rigorous impact studies on a selected sample of projects to verify data.

Resonance

The foundation is clearly not alone in the quest to assess the real impact of grassroots funding and to articulate that impact to decision makers. "*Hay sed*," observed a Venezuelan who works with international funding agencies—"people are thirsty" for ways to evaluate and communicate the effects of their work.

Other development assistance organizations that have consulted with IAF about the Grassroots Development Framework include the U.S. Agency for International Development (USAID), the Peace Corps, the African Development Foundation, the International Youth Foundation, the Bernard Van Leer Foundation, the World Wildlife Federation, and the newly formed

Andean Corporation for Grassroots Development. In conjunction with its matching fund agreement with IAF, the Venezuelan oil company PDVSA is introducing the framework as the basis for monitoring and reporting on its affiliates' contributions to grassroots development. As part of a cooperative agreement with IAF, the affiliates pledged to contribute $2.5 million to community development throughout the country in 1996. The prospect of a common framework and a common language energized the nineteen portfolio managers who attended a training seminar in August 1995. According to the managers, their greatest challenge is to present to headquarters, with credibility, "the majority of our results, which are on the intangible side."

In Colombia, a consortium of eleven foundations and NGOs recently opted to use the Grassroots Development Framework as its tool in common for documenting the impact on families, organizations, and communities of the jointly sponsored projects known as OCDs, or community development organizations.

The Ecuador ICS team is testing what its director called the "remarkable versatility" of the framework. Through an agreement with the recently created AIDS-prevention organization Alianza Internacional, COMUNIDEC is working with the local affiliate FIFSIDA, the Foundation for Initiatives Against AIDS. Seeking an alternative to "evaluations that count the number of condoms distributed," Alianza has engaged COMUNIDEC to adapt the framework to its needs and provide training for staff of the fourteen projects now under way in Quito.

"The key," reports COMUNIDEC's learning coordinator, "is the basic structure—the three levels, and the tangible and intangible sides"—which makes the construct adaptable to a wide array of programs. Evaluation in the health field, he finds, "tends to be ultra-technical, sophisticated, and expensive" and the province of specialists. In the framework, the front-line workers are finding "a tool that is accessible, that they themselves can use, that demystifies this business of evaluation." With the input of FIFSIDA, COMUNIDEC has adapted the framework—maintaining the basic structure but tailoring the titles of the categories and adjusting some of the variables to match specific goals. The organizational and societal levels remain essentially intact; the definition of direct benefits, particularly on the tangible side, changes to reflect results in prevention, treatment, and education.

The Lens You Look Through

The choice of lens that funders look through to assess the success of development programs matters mightily. In our bottom-line society, results tend to be equated with an immediate, tangible product—something that can be captured with a dollar sign or a snapshot. However, as veteran field-workers know, today's successful product often turns into tomorrow's white elephant, and the less visible efforts at building human capital and organizational clout are undervalued.

A recent book by Robert Putnam, director of the Center for International Affairs at Harvard University, gives historical weight and legitimacy to fieldworkers' empirical observations. In a series of essays and in his recent book (Putnam 1993, 1995a, 1995b), he makes a compelling case for what he terms "social capital," concluding that the historical record strongly suggests that the successful communities became rich because they were civic, not the other way around (Putnam 1995b). His conclusion is based on twenty years of meticulous and elaborately documented research on ten centuries of civic traditions in two distinct regions of Italy. He demonstrates that neighborhood associations, choral societies, and sports clubs are not just a nice by-product of economic prosperity; they are essential underpinnings of it.

> Citizens in civic communities . . . are prepared to act collectively to achieve their shared goals. Their counterparts in less civic regions more commonly assume the role of alienated and cynical supplicants. (Putnam 1993, 182)
>
> [Similar to] the notions of physical and human capital, the term "social capital" refers to features of social organisation—such as networks, norms, and trust—that increase a society's productive potential. Though largely neglected in discussions of public policy, social capital substantially enhances returns to investments in physical and human capital. . . . The social capital embodied in norms and networks of civic engagement seems to be a precondition for economic development as well as for effective government. . . . The implications for social and economic policy are far-reaching. (Putnam 1995a, 65–78)
>
> Those concerned with democracy and development . . . should lift their sights beyond instant results. Building social capital will not be easy, but it is the key to making democracy work. (Putnam 1993, 185)

Measures Matter

Putnam's findings take on particular significance in today's climate of development assistance, one that is characterized by fierce competition for development funds. There are two contradictory trends. In some spheres, funding is increasingly scarce. In others, large multilateral institutions, disillusioned with mega-programs, are looking for more positive results by channeling their resources through NGOs. In either case, the paradigm they work from—which determines the lens they look through—is critical.

Funders looking to downsize their portfolios stress efficiency as a criterion for rating and culling grant recipients. The efficiency they seek, however, often relates more to compliance or skill in meeting administrative demands than to effectiveness in fulfilling beneficiary needs or building social capital and leading to real change. Multilateral funders, for their part, see NGOs as potential surrogates for state agencies, to deliver goods and services and carry out public works more cheaply and expeditiously.

If the principal funders value and reward only the tangible, quantifiable results that are in vogue, then that is what NGOs (consciously or unconsciously) will emphasize. Cost and efficiency measures will produce bricks and mortar, but at the expense of the less visible, less marketable effort to build human capacity and social capital. We may kill the goose that is laying the golden eggs.

The Cone is a modest attempt to set things in a broader perspective. It is based on the recognition of the importance of physical, tangible results for beneficiaries but also places emphasis on results that strengthen the civic base of society and that, cumulatively, lead to change in practices, policies, and attitudes at higher levels. The hope is to provide organizations with the incentive and the tools to document and to disseminate their achievements in translating practice into policy.

Appendix 1: And the IAF?

In the course of workshops or presentations of the framework, a common question has been, what about applying the Cone to the foundation itself? How do the results stack up? The editors of this book had the same query. The following attempt at a response, unofficial and unscientific, is based on the author's personal reflections on the two decades of experience that culminated in the creation of the Cone and on the preliminary data that have been gathered so far.

It is an obvious question, but more complex than it appears at first blush. In the early days, it was commonly noted that "The foundation *is* what it funds." Likewise, in large measure, its results are reflected in the achievements (or shortcomings) of its grantees. Yet the foundation has worked with 3,000-plus grantees over twenty years and is by no means the only funder in the Latin American arena. There is, therefore, no tidy way of saying, "As a result of IAF's funding, X occurred."

Particularly at the first level of the framework—direct benefits to participants and their families—it is difficult at this stage to get beyond gross generalization. Clearly not all grants "worked." Some failed to produce the promised improvements in standard of living, and the majority of projects took longer than originally envisioned. However, most grants did produce results in the form of improved housing or health care or access to credit or education—those results often being significant in their local context. However, in the larger scheme of Latin American poverty, they represent but a drop in the proverbial bucket. The real measure of effectiveness of grants is not simply how many houses were constructed or how many loans were made and repaid, but whether they served as catalysts for innovation or broader application. The initial round of data collection indicated substantial contagion effect, or replication, but it is too early to draw hard conclusions.

Where the particular contribution of the foundation is perhaps more easily discernible is *personal capacity*, on the intangible side of the ledger. IAF early on recognized the importance of "cultural energy" as a driving force in grassroots development and broke important ground in its support of cultural expression and other creative endeavors to reaffirm identity and build confidence and self-esteem. Documentation of these experiences accounts for a significant proportion of the articles, books, videos, and films produced or funded by the foundation.

At the *organizational level*, the most obvious phenomenon over two decades is the proliferation of NGOs and grassroots organizations throughout the hemisphere. IAF is but one of many contributors to the rise of the NGO, but it was among the pioneer funders and has been one of the most consistent. Again, results are as varied as organizations funded. Particularly in the early years, NGOs' ability to capture funds often outstripped their capacity to administer them, and organizations wobbled under the weight of overfunding. However, the majority survived and (again, a gross generalization) improved their capacity to plan, administer personnel and resources, and form increasingly sophisticated linkages. As NGOs become institutionalized and, in some cases, distanced from the bases they serve, organizational issues center increasingly on the intangible side—vision, democratic practice, and autonomy.

At the third level, preliminary data point to results that have heretofore gone largely undocumented. This is the case particularly in terms of interaction of grantee organizations with public- and private-sector institutions and the degree to which many have influenced laws, policies, and practices. Agrarian law, environmental protection, social security regulation, and criteria for creditworthiness are some such arenas of practice and policy. Particularly interesting are the new partnerships springing up between the NGO sector and newly empowered (but chronically underfunded) municipal governments.

On a different plane, however, what about the foundation itself? How has it fared as an organization, and what effect has it had? Direct funding of NGOs—now taken for granted—was a novel idea when the foundation began funding in the early 1970s. Grassroots development, which became the foundation's trademark, has gained currency, in considerable measure through IAF's publications and the diaspora of its staff members and interns to other institutions. As an institution, the foundation's vision—of a responsive, nonbureaucratic government agency, unencumbered by domestic red tape and decoupled from short-term foreign policy—was positively revolutionary. Two decades before reinventing government came into vogue, IAF was a reinvented agency, lean on protocol and bold in commitment to its mission of service.

With time and the effects of a changing political climate, the courage of conviction dimmed. Presently, even as the foundation fosters institutional strengthening in Latin America, it struggles to preserve its own institutional identity, dwindling resource base, and autonomy. Although none of the efforts to abolish the foundation or to fold it into the mainstream foreign policy apparatus has succeeded so far, the insidious creep of bureaucracy did. An ever-increasing share of staff time and energy is directed to compliance with multilayered regulations, dulling spirit and sapping initiative.

Looking beyond internal issues, however, to the broader plane of development assistance, the very existence of IAF—its philosophy, its style, and its track record—has prompted change in the rules of the game. Little known and largely ignored in the early 1970s, NGOs now figure prominently in the development dialogue. Moreover, at least rhetorically, the multilateral funders now pay homage to the notions of participation, empowerment, and local "ownership" of process, for which IAF was an early and vocal proponent.

Less successful has been its ability to inform the terms of debate on foreign aid, either within Congress and the administration or with the U.S. public at large. Despite demonstrated success in crucial areas such as safeguarding the environment or dealing constructively with street children, the lessons learned by IAF have not translated into support for foreign assistance, even of the self-help variety. Social development, a cornerstone of IAF's charter, is out of favor domestically in the more pragmatic drive for the short-term, bottom-line, immediate return on the dollar.

Macro policy notwithstanding, the heart of IAF's mission endures—its locus, as ever, is in the field. It is in creative partnership with its ICS teams, local counterpart foundations, and grantees that IAF's current contribution lies. The modus operandi has evolved markedly, to keep pace with rapid change in the scope and sophistication of Latin American and Caribbean NGOs. Much of IAF's erstwhile role is now decentralized and carried out in-country. Ultimately, its impact will be gauged by the breadth and endurance of the ripple effect that can already be seen in the myriad replications of its grassroots funding philosophy and responsive style, and in the legacy of civic and service organizations it helped inspire, legitimate, and nurture.

Appendix 2: The Cone—Results Registration System

The following are the guidelines for field offices and grantee organizations for registering project outcomes. The purpose is to encourage consistency in the application of key variables, making them comparable for analytical purposes across different grantees.

- The Results System of the IAF is based on its Grassroots Development Framework, with its twenty-two variables and corresponding indicators.

- Six of these variables are considered key.

- The central database of the IAF registers a maximum of five indicators per variable. Of these indicators, two are standard (uniform for all projects), and up to three additional indicators are discretionary (freely chosen by the local support team or the grantee).

- The databases of each country maintain this information and any additional information that the in-country team considers important (and that the budget can cover).

Selection of Variables

When presenting a project for funding, the ICS team and/or the grantee selects the variables pertinent to the project.

Registration of Information

For each selected variable, the following information is registered:

- Departure point, or the initial status with respect to each indicator, recorded before the beginning of the project.

- Updating throughout the grant period, making use of normal monitoring visits (at least once a year).

- Post-project status two years after the termination of the grant.

The data to be recorded include:

- Unit of analysis.

- Unit of measure.

- Source(s) of information.

- Collection method(s).

- Examples.

- Comments on trends or context (to give perspective to the other, more quantitative data and to facilitate analysis).

- Observations about applicability of indicators to the project.

- Name of data collector and of person recording the information.

Information Sources

In order of preference, the possible information sources are:

- Beneficiaries.

- Grantees (implementing organization).

- ICS team.

- Key observers (persons close to the community who can opine on project results, such as teachers, extensionists, social workers).

- Third parties (persons and/or organizations outside the community who know the grantee and/or its beneficiaries and can opine on the project's results, such as another organization in the same network or a ministerial official).

The ICS is in charge of distilling the information and recording the most trustworthy data.

References

Avina, Jeffrey, with A. Lessik, A. Gomez, J. Butler, and D. Humphreys. 1990. Evaluating the Impact of Grassroots Development Funding. An Experimental Methodology Applied to Eight IAF Projects. Issues in Grassroots Development Monographs. Arlington, Va.: IAF.

Humphreys, Denise. 1994. Socio-political Performance Indicators for Grassroots Development Projects. An Experimental Approach Used in the Cases of Coopechayote, Costa Rica, and Central Cooperativa de Mineros, Bolivia. Paper presented to Study Group on Mediating Sustainability: Sustainable Agriculture and Rural Development, Wageningen University, Netherlands, November.

Putnam, Robert. 1993. *Making Democracy Work: Civic Traditions in Modern Italy*. Princeton, N.J.: Princeton University Press.

Putnam, R. 1995a. Bowling Alone: America's Declining Social Capital. *Journal of Democracy* (January): 65–78.

Putnam, R. 1995b. Social Capital. In People Centered Development Forum News-letter, 6 March, Column no. 76. Based on Putnam, R. 1993. The Prosperous Community: Social Capital and Economic Growth. *The American Prospect* 13 (Spring): 35–42.

Zaffaroni, Celia. 1997. *El Marco de Desarrollo de Base*. Montevideo, Uruguay: TRILCE/IAF/SADES.

4

Assessing the Merits of Participatory Development of Sustainable Agriculture

Experiences from Brazil and Central America

Irene Guijt

Over the past decade, the inadequacy of soil and water conservation (SWC) initiatives has stimulated a search for alternatives that has centered on participatory watershed development. These alternatives have sought to increase the role of farmers in the problem analysis and planning of SWC interventions rather than only in their implementation, as was common in the past. However, the process of seeking appropriate and effective forms of farmer participation is complex and time-consuming, requiring much skillful facilitation and devolution of power. These requirements present a series of organizational challenges, since planning, funding, implementation, and evaluation of activities need to be modified. The policy environment in which such changes take place also influences the viability of approaches that are more farmer centered. Will the effort that is needed to make this work be worthwhile and lead to more sustainable, environmentally sound, and socially inclusive forms of agriculture?

Assessing the merits of participatory approaches to SWC is difficult. The complexity of biophysical processes, the difficulty of assessing externalities, and the fuzziness of the notion of participation make monitoring and evaluation an arduous task. The range of factors that influence the success of a participatory process makes causal links tenuous at best. The choice of potential indicators presents a veritable minefield. How, then, to know what works?

This chapter describes three examples of innovative agricultural programs and how evaluation and monitoring fit into the search for more sustainable forms of agriculture. The first example concerns a government program for micro-watershed development in the southern Brazilian state of Santa Catarina. The second example is from Honduras and Guatemala and provides invaluable insights into impacts after program termination. The third example describes a new venture by nongovernmental organizations (NGOs) with rural trade unions and farming communities in the northeastern state of Paraiba, and how monitoring is embedded in the search for viable agricultural alternatives.

This chapter starts by discussing issues surrounding SWC and its evaluation. The three cases are described in terms of the project context, field process, and impacts to date. The conclusions offer lessons about two processes: the process of undertaking participatory watershed development for sustainable agriculture, and the process of evaluating its merits.

Soil and Water Conservation and Participatory Watershed Development

The definition of sustainable agriculture is elusive. As Campbell (1994, 386) says, "attempts to define sustainability miss the point that, like beauty, sustainability is in the eye of the beholder. . . . It is inevitable that assessments of relative sustainability are socially constructed, which is why there are so many definitions."

Questions arise as to what is being sustained and for how long, over what area, and according to which norms of success. Asking who will benefit and who will pay can reveal the paradoxes of sustainability. Dilemmas continue to face agriculturalists today: cultivating all potential cropland in the South (excluding China) would reduce permanent pastures and woodlands by 47 percent (FAO, cited in Gardner 1996). Should some areas be cultivated intensively, so that others can be taken out of production and remain pristine? In these situations, what is sustainable?

The pursuit, too, of agriculture is like trying to hit a moving target.[1] As policy environments change, knowledge grows and is destroyed, biological resources degrade and disasters strike, new employment opportunities arise, or constraining legislation is imposed, the goalposts are shifting, and strategies for sustainability move. What is sustainable for a time in one location may well prove untenable when conditions change. Agreement on criteria of sustainability remains contentious and makes the assessment of success a subjective undertaking despite attempts to achieve objective, scientific rigor.

However, despite differing definitions, most approaches to sustainable agriculture agree on the central role of the conservation of soil and water in some form (Reijntjes, Haverkort, and Waters-Bayer 1992; Pretty 1995; NRC 1989). This is, quite simply, due to the enormous extent of soil degradation, the burden on national economies, and, above all, the pivotal role that soil plays in crop production.

The Scale of Degradation

Since 1945, 11 percent of land worldwide has become moderately to severely degraded, equaling 1.2 billion hectares, approximately the combined surface area of China and India (Oldeman et al. 1991). Land degradation has, in some contexts, been hidden by practices that increase yields, notably increased use of inputs (Pla-Sentis 1992). In Brazil, for example, between 1983 and 1993 annual fertilizer use increased from 45 to 85 kilograms per hectare of cropland (WRI 1996). A United Nations study (Oldeman et al. 1991) argues that more than two-thirds of the abuse to soil is caused by agricultural and livestock production or the conversion of forests to cropland. Erosion is by far the most common type of land degradation, accounting for 84 percent of affected areas (ibid.). Agricultural mismanagement alone has damaged 38 percent of today's cropland (Gardner 1996). For example, in the Victor Graeff watershed in Rio Grande do Sul, Brazil, the clearing of forests and the clean tilling of agricultural fields reduced infiltration rates six times, to only 0.2 millimeters per hour, and erosion rates stood at 5.8 tons per acre (Busscher et al. 1996).

The loss of the resource base is generally being taken seriously by governments, as proved by the many external interventions the world over that have focused on the conservation of soil and water. Indeed, agricultural authorities have undertaken soil erosion control measures for over a century in some areas (Pretty and Shah 1994). One estimate of the total investment of the U.S. government in soil conservation alone since the 1930s is US$16 billion, with figures in some years exceeding $1 billion (Hudson 1991). However, despite this massive investment, soil loss in the United States is still estimated to cost the national economy $3 billion per year. The costs of SWC projects elsewhere are similar: US$300 million in Ethiopia (Hudson 1991) and US$1,900 per hectare in Peru (Treacy 1989). Taking the Peruvian example, with 500,000 hectare of terraces, 75 percent of which are abandoned and crumbling, this translates to a required investment in SWC totaling US$1 billion in Peru, or about 3.5 percent of gross domestic product (GDP) (Treacy 1989).[2]

Interventions have concentrated on physical structures, destocking, resettlement, and the fencing off of areas. These efforts to stem soil erosion and conserve water have been largely disappointing, and even counterproductive (Lutz, Pagiola, and Reiche 1994; Hudson 1992; Pretty and Shah 1994; Critchley, Reij, and Willcocks 1994; Reij, Turner, and Kuhlmann 1986; Magrath and Doolette, 1990). Technologies have been inappropriate, implementation of poor quality, and maintenance undervalued. The process of design, implementation, and maintenance has, in many cases, been based on the imposition of measures, external control, and inappropriate incentives. Usually, "programs were planned, designed and often mainly implemented by outsiders and there was insufficient involvement of the farming community" (Hudson 1992, 74).

Fundamental Shifts

Progress has been made in redefining SWC. For example, there is increasing recognition that:

- Engineering-based soil conservation deals with symptoms and not with causes of land degradation.

- Imposed SWC programs rarely have a lasting impact.

- Soil conservation is not usually perceived by rural people as their most pressing problem, but decreasing productivity often is.

The alternative approaches[3] to participatory watershed development, or "total catchment management" (Martin 1991), that are taking root are characterized by three significant differences:

1. Adopting the (micro) watershed as the unit of analysis and intervention.

2. Placing local men and women at the center of the interventions.

3. Focusing on SWC as the interaction of social and biophysical systems.

Consciously creating opportunities for farmers to have a say in the design, implementation, maintenance, and evaluation of catchment planning is perhaps the most significant transformation in SWC. But all three features are closely related. Adopting a watershed perspective creates the imperative to work with groups of farmers or communities in a coordinated manner. And working with farmers requires a solid understanding of the social processes in which their agricultural activities are inserted (Cornwall, Guijt, and Welbourn 1993). It means paying attention to both local and external institutions and social structures that design, implement, and manage the interventions. This, in turn, calls for an intersectoral approach to catchment management and policy changes (IDB 1995; Ostberg and Christiannson 1993; Thompson 1995; Pretty, Thompson, and Kiara 1995).

However, although there is considerable anecdotal evidence that the surge of participatory watershed programs is having positive social, environmental, and economic impacts, there is a lack of extensive and systematic data. In particular, there is little clarity about the nature of farmers' participation in these experiences and of the impact this has had on success rates. Yet if the small-scale accomplishments made to date are to spread, it is essential to understand the factors of success. Equally important is the need to understand the constraints for success and replication. Monitoring and evaluation (M&E) processes are the logical route to seeking this information, yet they are often inadequate. Data are often simply not available, even for conventional SWC activities. Hudson's study (1991) of the reasons for success of soil conservation projects found not a single agency that had sufficient documented information to enable project evaluation.

The Learning Imperative

The monitoring of data also plays a crucial role in changing policies that block the spread and sustainability of small-scale successes (Guijt 1996; Pretty 1995). When advocating policy changes within organizations or government agencies, the absence of data, or proof, becomes even more debilitating. Such proof can come only by strengthening the learning within SWC programs and sharing insights into the processes and outcomes. Evaluation methods are required that allow organizational reflection and learning from rural people, yet these are sadly lacking in most conventional approaches to evaluation (Korten 1980; Chambers 1992; Pretty 1995; Thompson 1995).

Conventional Approaches to the Evaluation of Soil and Water Conservation

Conventional evaluation approaches to SWC view their object largely as a biophysical process. As they are not imbued with the principles of participatory watershed development, they ignore sociopolitical and institutional processes. Not surprisingly, evaluation in this context has often been limited to the counting of outputs or the efficiency of implementation (Hudson 1991). Where the ecological and economic impacts have been assessed, these have been limited to a restricted geographical zone, often a hydrological unit (Bishop 1992). Impact studies have tended to focus on erosion-productivity links, but these are notoriously difficult to establish.

Three common conventional evaluation approaches applied to the SWC sector are (de Graaff 1993; IDB 1995):

1. With/without comparison, which compares the production or productivity of areas with SWC measures to that of areas without such measures.

2. Measurement of the economic benefits of downstream effects in with SWC and without SWC cases (compare IDB 1995).

3. Multicriteria analysis, which allows variables or criteria to be expressed in their own units, rather than forced into a common quantitative unit, resulting in a type of pair-wise comparison of the different criteria.

These three approaches require large amounts of quantitative data and considerable economic expertise. As this will inevitably exclude non-economists, these approaches are of limited value in most situations of self-evaluation. The object of valuation is generally determined by outsiders (often economists) and must generally be quantifiable. Only a narrow spectrum of causal relationships is assessed, often based on tenuous assumptions. Qualitative criteria, such as increased local management capacity or farmers' increased independence from monopoly markets, play no role in such assessments. These approaches focus on the biophysical aspects of SWC and do not usually include the social structures and institutions that sustain measures.

Conventional evaluations are inadequate for assessing participatory watershed programs, for the following reasons:

1. Objectives are unhelpful for local learning, focusing instead on accountability needs of external actors (such as funders).

2. Timing is inappropriate, as evaluations are not carried out regularly, thus missing opportunities to adjust activities.

3. Inadequate indicators focus on short-term biophysical measures and some operational aspects,[4] excluding the process of community participation.

4. The wrong implementers are used, with evaluation carried out by donors or donor-initiated external experts, thus limiting the extent of local learning.

5. Exclusive methodologies are based on written forms and questionnaires, which are not conducive to the inclusion of farmers in the crucial analysis stage.

6. There is limited feedback of evaluation findings and related decisions, thus hindering the spread of essential lessons.

Few SWC projects emphasize farmer-centered processes as essential for project success. Nor do they emphasize this in evaluations. In conventional project implementation and evaluation alike, the product is the focus and is divorced from the process. Assessing participatory watershed development requires paying attention to the process itself, and that is best carried out with those who are involved. The experiences from Latin America presented here are rooted in this principle: the necessity for social processes of transformation, be they individual or collective, in resource management. SWC is embraced as an integral element of sustainable agriculture and rural development strategy, rather than as a separate, technical element in a policy or resource management strategy. To further the cause of sustained SWC, evaluation processes must be consistent with this principle.

Participatory Evaluation of Watershed Development

Self-evaluation and participatory monitoring (participatory monitoring and evaluation [PME] and participatory impact monitoring [PIM]) are alternative approaches to conventional evaluation.[5] Their point of departure is improving the internal learning process and the planning and implementation of interventions, whether by farmers, NGOs, or government agencies. Evaluation in this context means emphasizing its value as a tool for organizational learning, rather than the control and accountability it is usually associated with. Monitoring becomes an opportunity to reflect and adjust.

There is limited experience and even less documentation of self-monitoring and evaluation of agricultural and rural development experiences[6] from Sri Lanka, India, Brazil, and Senegal (Uphoff 1992; Sommer 1993; Shah and Shah 1994; Sidersky 1995; A. de Groot personal communication; B. Gueye personal communication; Rugh 1986; Davies 1996). But recent studies of some on-farm research programs, farming systems research projects, and mainly government SWC interventions in a wide range of countries have found that learning was generally weak (Hudson 1991; Merrill-Sands et al. 1991; Pretty 1995).

Self-monitoring experiences are fundamentally different from conventional approaches to monitoring. If the object of evaluation changes, such as a shift from erosion control to participatory watershed development, evaluation approaches need to be adjusted accordingly. Translating this to the SWC sector means redesigning the monitoring process to take into account both the sociopolitical and the biophysical systems in which measures are nested. It means involving stakeholders in monitoring and evaluation design and analysis and seeking a wider range of indicators of success beyond those of productivity and externalities (compare IIED 1992). It calls for ensuring that findings are passed on to those who have been involved and can benefit from them. However, participatory monitoring and evaluation of participatory processes face many practical and conceptual challenges, partly due to the complexity and diverse interpretations of the term "participation."[7]

Evaluating Participatory Watershed Development in Latin America

The following sections summarize the work undertaken by current or former staff of three programs aiming to support more sustainable forms of agriculture, with a strong focus on soil regeneration. The first two studies assess agricultural programs that were developed in close collaboration with farmers, but neither of which emphasized farmer-based monitoring as fundamental to the program. The third example highlights how participatory monitoring is embedded in a farmer-centered approach to agricultural development.

The EPAGRI Program of Microcatchment Development[8]

Project Context. Santa Catarina, located in southern Brazil, is a small state by Brazilian standards, covering about 96,000 square kilometers. It is a state of small farms. Of the 235,000 farming households, 89 percent have less than 50 hectares, and 40 percent of these are less than 10 hectares. Fodder maize covers about half the area of annual crops.

Soil preparation in Santa Catarina is conventional, with much disturbance to soil by ploughing and harrowing, aggravated by increasing mechanization. This has affected soil fertility, and crop productivity has decreased over time, now averaging around 3,200 kilograms per hectare. Despite small holdings and erosion-sensitive soil, it is the fifth largest food-producing state in garlic, apples, pigs, poultry, beans (first harvest), tobacco, and onions.

For the past ten years, EPAGRI (State Agricultural Research and Extension Agency) has pioneered a microcatchment approach. In 1984, its efforts to control erosion increased, sparked by two years of heavy floods. It adopted the microcatchment as the planning unit for natural resource management, following earlier experiences in the neighboring state of Parana. After two years of small-scale initiatives in fourteen municipalities, a national-level program for river microcatchments (PNMH) was created in 1986 by the Ministry of Agriculture. This pushed up the number of participating municipalities to sixty-eight, but it was only with International Bank for Reconstruction and Development (IBRD) funding (approved in 1991) that large-scale efforts were made possible. Over a seven-year period, 80,000 producers in 520 microcatchments are to be involved in the work.

EPAGRI's microcatchment program aims to increase the labor productivity and net profit of farmers by regenerating and conserving the productive capacity of the soil and controlling pollution in rural areas (particularly waterways contaminated by piggery effluents). Soil conservation is key in the strategy, with activities focusing on increasing vegetation cover, controlling surface runoff, and improving soil structure. Improving soil fertility with green manures plays an increasingly important role.

Field Process. Although the social, economic, and biophysical uniqueness of each microcatchment influences the final process of watershed development, most programs follow the series of general steps outlined below, based on the timing of various stages in one of the older catchments of EPAGRI's work, Riberão das Pedras.

1. Preparation. EPAGRI senior technical staff organize a municipal-level meeting with the key local leaders, entrepreneurs, bankers, and technical experts. This usually includes a tour of another active microcatchment. The group then identifies the microcatchments in the municipality and nominates four to be invited to join the catchment program, based solely on technical criteria. The nominated catchments are put before the municipal council, with farmer representatives, for final validation. Only then does the field-level work start.

2. Microcatchment Motivation. EPAGRI considers this the most important phase. The extensionist responsible for the microcatchment visits all the

households for initial discussions. This also includes a community-level meeting with leaders or families. The extensionist organizes a tour for about forty families to visit an active microcatchment, a farm with an advanced soil management approach, or EPAGRI's training center or research stations[9] to look at green manure units and soil preparation systems. Only then do the families in the nominated catchment vote whether to commit themselves to the microcatchment work. Work will proceed only if at least 70 percent of the farmers of each microcatchment consent to the partnership. During this phase, support is provided by EPAGRI's regional technical advisers.

3. Planning of Activities. SWC activities are planned for a four-year period. An individual property plan, or PIP, is prepared for each farm, generally by the extension agent, based on discussions with each farmer (almost always a man). The PIP describes present land use, natural aptitude, potential use, and conservation plans, in line with the agricultural objectives of each farmer. Activities that are eligible for financial support from PROSOLO, a state-run one-year grant system with strict conditions, are also identified.

4. Implementation. The conservation measures are implemented with the extension agent, usually only upon the release of funds from PROSOLO. In most but not all microcatchments, catchment committees are elected to guide the overall implementation of activities, particularly those that require collective action and support. Individual activities include piggery construction, toxic waste depot construction, use of green manure seeds, tree nursery development, manure application, and calcium application. Collective activities include the construction of a water tank, a toxic waste depot, or a maize storage facility or the hiring of various machines, such as a direct seeder, a manure spreader, or a bulldozer.

Impacts to Date.   Between 1991 and 1994, EPAGRI staff reached over 38,000 farmers in more than 255 microcatchments. Over 11,000 PIPs were developed, more than 4,300 tonnes of green manure seed distributed, about 1,540 piggeries constructed, and over 1,800 spring protection works completed. Research from two catchments showed that the increase in value over five years, from only three key agricultural products, was more than the total investment of PROSOLO and of individual farmers over the five-year period (Freitas personal communication).

There is no farmer-based monitoring system in place. Instead, EPAGRI produces quarterly reports on the microcatchment program, largely for external accountability to donors. These reports identify fourteen types of quantitative data, such as the number of new participating farmers; microcatchment and property plans completed; motivational meetings, excursions, and training courses held; individual activities (per activity type, of which there are fourteen); collective activities (per activity type, of which there are ten); improved roads; water quality samples analyzed; and basic infrastructure constructed (schools, protected wells, latrines, rubbish dumps). Other,

Qualitative Evaluation of PROSOLO Performance

■ The execution of individual activities was good in 86 percent of the cases, medium in 23 percent, and bad in only 1 percent.

■ If PROSOLO did not exist, only 50 percent of the individual property measures would be implemented, and only 15 percent of the collective activities would be implemented.

■ Sixty-two percent of farmers implement individual measures only after receiving PROSOLO funds.

■ Seventy-one percent of farmers are satisfied with their PIPs, and 79 percent are satisfied with the technical assistance they receive in the implementation of their property plans.

■ There is great delay in the PROSOLO funding process: three months to develop a proposal, four months to contract the proposal, and another one and a half months to receive the funds.

Source: J. M. Paul. 1994. Avaliaçao Qualitativa do PROSOLO/Microbacias. Resumo das Principais Conclusoes e Recomendacoes. Mimeo. Instituto CEPA/SC.

more qualitative impacts of the program are also documented, which provide recommendations for particular parts of the program, such as PROSOLO.

Although this type of data obviously provides information on the quantitative objectives of the program, it does not provide insight into the impacts and sustained value for local people, the local environment, and the economy. There is also no comparison with those farmers not participating in the program. Thus it is impossible to assess whether participation in watershed development is worth the effort and what the likelihood of sustained impact will be. As part of the New Horizons program, other impacts within the microcatchments were analyzed to better understand issues related to the sustainability of agricultural development.[10] Broader impacts were noted for the following two areas (Freitas 1995; Guijt 1994):

Extensionist's Role and Identity.   The extension agent is the key link in the participatory watershed program. The greatest difference with past EPA-GRI work is that extension agents now concentrate their efforts in a well-defined area. As this allows for more contact with farming households and better rapport, job satisfaction is higher. The PIP process means that extension agents understand local agricultural realities in great detail, thus gaining the respect of the farmers. Results are faster and more clearly a result of the

Impacts at Microcatchment Level
in São Lourenco do Oeste

The work in the municipality of São Lourenco do Oeste started in 1985 in the microcatchment of Rio Macaco, with fifty families, about 1,800 hectares, and 1,800 millimeters of rainfall per year. The livelihoods of the families were based on a few agricultural products—namely, maize, beans, and pigs. The agricultural fields covered very steep areas, based on a conventional system of tillage with ploughs. There was little use of green manures and no conservation measures.

With the implementation of the microcatchment project came a large change in the adoption of conservation measures. In 1985, there were already 8 kilometers of terraces and contour grass strips. There was much experimentation with green manures. Today, more than 70 percent of the area is planted in winter with green manures and there are 40 kilometers of terraces. Direct planting is increasing and now covers 10 percent of the cultivated area. Productivity of maize has increased by 49 percent among farmers adopting measures. There has also been an enormous diversification in agricultural activities to include vegetable growing, fruticulture, maté tea, and increased production of milk cows and pigs.

Soil degradation has largely stopped and river pollution has been reduced due to the construction of piggeries and pig stalls and the decrease in horseflies. Storm flood levels have stabilized as a result of the conservation measures. A rainfall of 70 millimeters in January 1993 still raised the level by 1.9 meters. In May 1994, a rainfall of 102 millimeters raised the level of the river by only 40 centimeters.

Elsewhere, in Riberão das Pedras, a no-tillage system is used on 80 percent of the cultivated area, cutting all soil disturbance. Maize production has increased by 60 percent in ten years. Terraces were almost completely abandoned. Instead, 80 percent of the cultivated area is now covered in winter with green manure crops (mainly oats and vica).

Source: V. H. de Freitas. 1995. Transformations in the Micro-catchments of Ribeirão das Pedras/Agrolândia and of Rio Macacos/São Lourenco do Oeste, Santa Catarina—Brazil. Paper presented at New Horizons, IIED international workshop, Bangalore, India, December.

efforts of the extension agent. Both farmers and municipal authorities appreciate the role of the extension agent more. This has led to 101 of the 192 extension agents becoming direct employees of the municipality. Recently, extension workers employed by EPAGRI have started identifying themselves as part of the prefecture team. As EPAGRI staff can commit themselves to only two years per microcatchment, this handing over of staff to local institutions will enhance sustainability of the catchment approach. Support is still provided by sixteen regional level EPAGRI agronomists.

Innovative Partnerships.   Scaling up of activities has meant easier adoption in subsequent catchments due to off-site awareness of program activities. But as extension staff keep a maintenance role in all catchments, they will soon have limited time to invest in new catchments. Innovative partnerships are essential to ensure that the work spreads.

As the microcatchment project has always been conceived as being the responsibility of the whole society, support is actively sought from the non-state sector. To this end, conscious efforts have been made to create linkages with the private sector, academic institutions, and municipal departments:

■ Students and faculties from agricultural colleges are participating in the elaboration of PIPs through paid internships.

■ Farmers are being trained to act as promoters at the microcatchment level, taking over the task of the extension agent. EPAGRI's training program has created a soil conservation, use, and management course for farmers to support this process.

■ There are fascinating links with the private sector. The tobacco and poultry and pig industries (SINDIFUMO and SINDICARNE) are realizing that they can sustain their industries only if soil fertility is maintained. Chicken and pigs eat maize, which must be grown sustainably, as must tobacco. EPAGRI has trained many staff employed by the two industries in soil management, and they are training farmers. Also, SINDIFUMO and SINDICARNE now pay for about fifty of EPAGRI's extension staff to support the microcatchment teams.

Facing New Challenges.   EPAGRI recognizes the many challenges it faces to strengthen the work, improve the quality of the process and the outcome, and spread the impact. Questions it is asking include:

■ How can the PIP process be more participatory? There is high dependency on external agents to measure and plan.

■ Can the involvement of external agents be phased out, or is this a pipedream? This has not been thought through, despite the limitation of two years in each microcatchment.

■ How can women farmers and professionals be more involved? Women often make an essential contribution in the adoption of measures, but very few women participate in exchange visits or training, and there are none on the catchment committees. There are no gender-differentiated data, other than one figure: three of the 192 EPAGRI extension staff are women.

■ How can participatory monitoring be institutionalized? So far, only a few quantitative impacts are being analyzed in a handful of catchments. The New Horizons–type evaluation has not been repeated.

World Neighbors in Honduras and Guatemala

In 1994, a study was carried out by the Honduran organization COSECHA (Association of Advisers for a Sustainable, Ecological, and People-Centered Agriculture) to look at long-term impacts of sustainable agriculture programs (Bunch and López 1995). This section is a summary of three programs in Guatemala and Honduras that highlight impacts up to fifteen years after the termination of outside intervention.[11] Their version of farmer participation focused on collaborative development of locally suitable low-external-input technologies through farmer experimentation and mechanisms to share and spread these technologies among other farmers.

Project Context. The Cantarranas Area. The hillsides around the central Honduran town of Cantarranas vary in slope, with an average rise of about 30 percent for each 100 meters. The forests have been seriously degraded. The climate of the Cantarranas area varies from hot and semiarid, with frequent and severe droughts in the bottom of the valley, to a cool climate, with sufficient rainfall for six months each year. Between 1987 and 1991, the Cantarranas Integrated Agricultural Development Program, financed by Catholic Relief Services and managed by World Neighbors (WN), worked in thirty-five villages (Bunch 1990). The program worked almost entirely with small farmers with two- to five-hectare landholdings. Using in-row tillage and intercropped green manures as its starting point, it expanded into a general program of agricultural development and preventive health measures.

The Guinope Area. The Guinope area has similar variations in altitude and rainfall as Cantarranas, but with less severe slopes. An impenetrable subsoil beneath the thin topsoil prohibits agriculture once it has eroded. Before 1981, emigration from the Guinope area was heavy. Between 1981 and 1989, a similar WN program worked in forty-one villages in southeastern Honduras (Bunch 1988). This program also focused on soil recuperation, basic grain production, crop diversification, and preventive health measures. The program's main technologies were drainage ditches with live barriers and the use of chicken manure.

The San Martin Jilotepeque Area. The San Martin municipality lies 50 kilometers west of Guatemala City. The southern half of the municipality, where the program worked in forty-five villages, varies in altitude from about 800 to 2,000 meters and has enough rainfall for a good maize crop most years. The mainly Cakchiquel Indian population had little land, owning on average less than half a hectare of seriously degraded land per family. The San Martin Integrated Development Program in Guatemala was financed by Oxfam UK and carried out by WN between 1972 and 1979 (Bunch 1977). It was an integrated program, dealing with agriculture, health, road construction, functional literacy, cooperative organization and so forth. The program used contour ditches and a side-dressing of nitrogen on maize as the initial technologies to motivate people.

Impacts to Date.   The impact study encompassed twelve villages and about 900 farmers. The villages per region were selected by COSECHA staff as follows: one village in which they judged the impact to have been best, two villages in which the impact was moderate, and one village in which the impact was relatively poor.

The discussions looked at changes in range of indicators over time: from program initiation to program termination and on to the time of the study (1994). Quantitative indicators included the numbers of farmers using regenerating technologies and maize and bean productivity (see Tables 4.1 and 4.2).

These results show that the overall level of continuing innovation, despite program termination, is high. Soil conservation and soil improvement have continued. The lack of maize production in Guacamayas is because farmers earn more money from vegetables and prefer to buy maize. Thus, even the lack of maize cultivation is due to large local increases in yields and value of production of horticultural produce. Other positive impacts in the project areas include (compare Bunch and López 1995):

- Increased wage rates, land value, number of trees planted, crop diversity, and practice of intercropping.

- Decreasing or reversed emigration from the project areas and less resource degradation.

- Virtual halt of the use of herbicides through hand weeding or use of green manures.

- An increase in local savings, leading to decreased dependence on formal credit and increased investment in education, land improvement, purchase of animals, and so forth.

Less success has been achieved in reducing farmers' dependence on insecticides and fungicides. Central America, in general, has been slow to find feasible alternatives for these chemicals (Bentley and Andrews 1996). The most disappointing finding is the lack of spontaneous technological spread *between* villages, despite the existence of spontaneous spread *within* villages.

The overall level of continuing innovation in the study sites, despite program termination, has been remarkable. In San Martin, over thirty innovations have been adopted successfully since program termination. Probably most important is that each village has developed at least one completely

Table 4.1 Number of Farmers Using Other Technologies at Project Initiation, Project Termination, and Time of the Study

Technology	San Martin	Guinope	Cantarranas	Total
Contour grass barriers				
IN	1	0	0	1
TE	<100	44	48	<192
94	203	33	44	280
Contour or drainage ditches				
IN	1	0	0	1
TE	136	56	39	231
94	162	43	34	239
Green manures				
IN	0	0	0	0
TE	21	0	14	35
94	38	2	12	52
Crop rotation				
IN	0	12	0	12
TE	6	56	<97	<159
94	10	125	119	254
Fields no longer burned				
IN	na	2	0	2
TE	na	83	77	160
94	na	127	108	235
Fertilization with organic matter				
IN	<10	4	30	<44
TE	<42	100	53	<195
94	>124	213	60	>397

Note: IN, project initiation; TE, project termination; 94, time of the study. These figures represent approximate data, as the fifteen-year-old end-of-program reports for one village in San Martin and one in Guinope were not complete. The total values for these rows are therefore also estimates.
Source: Bunch and López 1995, 7.

new system of production. These innovations include the cultivation of new crops, agroprocessing, forest management, livestock management, and the use of composting latrines.

Of particular interest, and related to rates of innovation, is that of the dozens of technologies promoted by the various programs, only one technology, crop rotation, has survived in its original form for at least fifteen years. Bunch and López (1995) argue that this lack of technological sustainability is probably not caused by poor selection of technologies.

Table 4.2 Productivity of Maize at Project Initiation, at Project Termination, and at the Time of the Study (in 100 kg/ha)

	San Martin	Guinope	Cantarranas
IN	16	18	20
TE	<100	72	39
94	180	82	41

Note: IN, project initiation; TE, project termination; 94, time of the study.
Source: Bunch and López 1995, 8.

Instead, they say that most of the technologies have become obsolete as changing local circumstances have reduced or eliminated their usefulness. Former program staff describe one such situation:

> When the grass barriers trap eroding soil and build up a natural terrace (some four to six years after adoption), farmers stop cleaning out their contour ditches. As many villagers have explained, "if my ditches never fill up with water any more, why should I keep cleaning them out?" Both in-row tillage and cover crops can make contour barriers irrelevant.

Bunch and López's study highlights that aiming to ensure the sustainability of specific technologies may be barking up the wrong tree, and may even be counterproductive. Much more relevant to farmers' well-being and productivity seems to be aiming to sustain the process of innovation. What is essential is that productivity continues to climb as a result of continuous local-level innovation.[12]

If the technologies themselves have a short life span, development efforts with sustainable agriculture must focus on local-level innovation. For this, farmers need to:

■ Learn the basics of scientific experimentation.

■ Learn basic theoretical ideas about soil and agriculture to focus their experiments on relevant areas.

■ Learn to share the results of their experiments with one another.

■ Be motivated to do all of the above on a long-term basis.

AS-PTA Nordeste and the Rural Trade Unions

Project Context. AS-PTA is a Brazilian NGO that has been working with agroecology, family agriculture, and sustainable development since 1983. It strives to identify and promote more sustainable rural development approaches than those currently adopted in Brazil. Its activities focus on field research and the extension of appropriate technology for small-scale producers, networking,[13] and advocacy.

To achieve its goal, AS-PTA follows a strategy based on the implementation and analysis of practical examples for awareness-raising. Alternative development routes, it argues, can only be pursued once their feasibility has been demonstrated in concrete ways at the farm and microregional levels. This requires social, technical, and institutional transformation to support the local-level improvements. The fieldwork that AS-PTA engages in with rural trade unions and farmer associations serves as a testing ground for analytical processes and agricultural practices that are then offered to others.

AS-PTA's Nordeste program works in Paraiba, one of Brazil's driest states. The fieldwork is focused in the Agreste, a transitional zone between the semiarid interior and the humid coastline. This zone is characterized by its relatively high population density (20.4 percent of the population on only 9.5 percent of the area) (Andrade 1980) and its environmental diversity. Average rainfall figures can vary between 350 and 850 millimeters per year within one catchment. Although a range of crops is grown, agriculture focuses on maize, beans, and cassava. Sweet potato, banana, potato, and fava are also common. Small-scale livestock is an important supplement to diets and incomes.

AS-PTA's field activities focus on the municipalities of Solanea and Remigio, which cover 900 square kilometers, 50,000 inhabitants, and 4,000 smallholders (Sidersky 1995). The project activities are carried out by a team of five agricultural professionals in partnership with *animadores* (motivators), who are active members of the municipal rural trade unions. These unions, the STRs, carry on the agricultural activities once AS-PTA moves on to other municipalities, so they are central to the capacity-building work. Besides these two membership organizations, the collaboration also includes community associations and budding farmers' interest groups. The latter have no formal structure as yet, but they meet to discuss specific agricultural innovations, such as integrated pest management in banana stands and pigeon pea intercropping. Loose contacts are maintained with the local university, EMATER (the state-level extension service), and other community development NGOs.

Field Process. Unlike EPAGRI, AS-PTA has no set pattern of work, as its efforts are recent and have been limited to two municipalities. The field process has consisted of a series of steps focusing on capacity building and institutional support; participatory problem identification, monitoring, and evaluation; extension of technological alternatives; and farmer experimentation.

Initial efforts were invested in setting up a collaborative process with the two STRs that are part of a strong national movement in Brazil. The focus of STR activities varies by region, depending on which local issue is most urgent: the threatened viability of small producers, the working conditions of agricultural laborers, or the political struggle for land rights in areas of evictions and insecure land rights. Many STR representatives in Solanea and Remigio are keen to work with AS-PTA, as they realize that the agricultural development alternative is a valuable, more practical complement to their existing work with political rights.

The first stage of the work, starting in September 1993, involved a six-month exploratory phase. The situation analysis using the participatory agroecosystem appraisal (DRPA) with thirty farmers and STR representatives from the two municipalities centered on the agricultural crisis of small producers and ways in which local organizations could confront this. This process described distinct microzones and led to a program of prioritized activities for each microzone. The analysis was disseminated to about forty communities (1,450 farmers) using slide shows developed with and facilitated by STR.

From 1994 onward, the partnership has pursued a permanent planning process alongside the implementation of identified experimentation and extension activities. Planning seminars are held annually with about forty farmers to address the joint challenge of environmental regeneration and economic viability of smallholder production. Outputs from monitoring and annual evaluations provide essential planning inputs. Current field activities focus on farmer-to-farmer extension, participatory technology development, and institution building.

Concrete activities center around increasing low crop production and low productivity caused by soil degradation and lack of seed, through community-based seed banks and contour planting; fodder provision in the dry summers, via pigeon pea intercropping; and improved banana production (a significant cash crop) via biological pest management. Alongside these are many farmer experiments covering seven themes, ranging from ant control to green manuring of bananas.

AS-PTA is careful to distinguish three levels of farmer participation in its work:

1. A nucleus of about ten farmers, affiliated with the STR, in charge of strategic planning, data analysis, monitoring, and evaluation. This group is also responsible for most dissemination and monitoring work in the field, and its members are the *animadores*.

2. Between sixty and eighty farmers, men and women, including community association leaders and individual farmers engaged in joint experimentation. Practically all these farmers are also involved in key moments of monitoring, evaluation, and planning, particularly those related to the experiments.

3. Activity-specific collaboration with the general farming public and community associations, covering thirty communities and over 500 farmers, that
are keen to adopt particular measures.

The methods used include training for agricultural instructors (all of
whom are farmers); audiovisuals to provoke discussion on agricultural
development alternatives; evening sessions with innovative farmers as
"guest lecturers," providing their own experiences and slides discussing
each innovation; and farmer-to-farmer visits. Courses have been held on
agroecology and the use of municipal budgets for agricultural development.
The use of thematic DRPA continues to deepen understanding and lead to
new experiments. In all these activities, STR affiliates have played a key
role in design and facilitation, taking on increasing responsibility as skills
have grown.

Impacts to Date.   Although AS-PTA's work is too recent to allow a
detailed assessment of its impact, a few highlights indicate some trends. It
has been approached by a third STR in a nearby municipality wishing to
enter into the partnership. The number of farmers taking on experimentation has grown from ten to about seventy-five, organized in eight groups.
Work reaches at least 500 farmers in thirty communities directly, as well as
an unknown number of farmers who have learned by replication.

Of particular interest is the urgency with which AS-PTA is pursuing the
issue of monitoring. It is integrating participatory monitoring and appraisal
processes seriously from the onset of the collaboration, as an essential part
of its capacity-building efforts. A long-term collaboration[14] is under way to
develop, with the STRs and farmer interest groups, a participatory monitoring
system with two main objectives:

1. To provide an ongoing learning experience that can help strengthen group
structures and improve the planning process and relevance of the interventions.

2. To provide data to fulfill accountability conditions of donors and to support local and national policy efforts of the STRs and AS-PTA alike.

AS-PTA has developed a basic M&E form for farmers and STR representatives to help evaluate the extent of adoption and success of new activities.
Data collected focus on numbers of farmers involved in discussion and numbers of meetings, workshops, experiments, and so forth. Yield data are being
collected, but accurate documentation is proving difficult for the STR monitors. Evaluation of the data is complex, as one farmer may use several new
technologies at once. There is no agricultural baseline data in the region, further complicating the monitoring of progress.

Methodological Challenges of Participatory Monitoring. Although the process has started only recently, many questions have already been raised about the nature of participatory monitoring of sustainable agriculture, which is embedded in a highly political process of strengthening trade unions and other farmer organizations (Guijt and Sidersky 1996).

■ What is monitored and evaluated—only the product, or also the process? AS-PTA and the STRs see a clear need for hard data of tangible benefits. Yet how can this be obtained effectively and rigorously without forgetting the need to assess the process of institution strengthening and the independence of farmer innovation?

■ How can one deal with the multiple agendas that arise through such partnerships? Who is to monitor and evaluate which indicators if farmers, STR, and AS-PTA have different priorities? What extra work does it mean for AS-PTA, farmers, and STR, and who foots the bill? The farmers' area of operation is smaller than the area covered by a rural trade union, not to mention AS-PTA. Each has unique preferences for specific indicators, so negotiating these and the division of responsibility is essential. The definition of objectives and their prioritization has already taken six months and about ten meetings. Each trade union has had extensive internal negotiations to resolve the tensions, highlighted by the monitoring process, between its political aims and the practical realities of its collaboration with AS-PTA. The first workshops in January and July 1996 helped identify:

1. The objectives of the partnership according to two central stakeholder groups (AS-PTA and STR affiliates), plus priority areas for monitoring.

2. The indicators, that is, information they need in order to assess whether they are achieving their objectives.

3. The best methods to collect and register the information, with several innovations to suit the desired indicators and the local cultural context (Guijt and Sidersky 1996).

■ Subsequent steps include further development of monitoring processes with farmer experimentation groups, further adjustment of methods of data collection and data analysis and the monitoring methodology, and dissemination of the monitoring methodology to other NGOs and trade unions in Brazil. Of central importance are the usefulness and end use of the data collected and the long-term sustainability of the monitoring process for farmers and trade unions.

■ How can one evaluate in ways that are inclusive of the various stakeholders and that build local capacities, allowing the integration of data collection into daily working practice? Inclusion of stakeholders in the design and implementation, but above all in the analysis and adaptation, of the monitoring methodology is helping to ensure that local capacities are built and that the methodology is realistic.

■ What level, or unit, of analysis is best? This influences the choice of both indicator and method and requires clarification soon. In many situations, community-based change would perhaps be the most obvious unit to monitor. However, in Brazil, a cohesive community that identifies itself as such is difficult to find. In most cases, a community is simply a cluster of houses. People's identities might relate to the local chapel or the local school, not to a geographical area. When monitoring objectives—for example, a growth of 25 percent in fennel-producing farmers—it is necessary to know at which level the change is found. Does this mean growth at the "community" level within an association, the municipality, the region, or the country? Currently, AS-PTA and the STRs are analyzing various units, as they have objectives for different levels: the associations (if they are organized as such), the municipalities (indicators at the level of a syndicate), special farmer interest groups (connected to a specific activity), and the family/farm.

■ How does one decide on key benchmarks? To make a comparison over the long term, it is necessary to have an initial basis of information. However, on what basis can these benchmark indicators be chosen? How can one account for the possibility that impacts are technically complex, such as "slowing rates of degradation"? Clearly, indicator choice is strongly related to objectives. There is no point in assessing a benchmark indicator related to an objective that has not been set for the future. But with the lack of agreement on and the changing nature of the objectives of sustainable agriculture, benchmark indicators require difficult subjective choices.

■ How does one deal with external influences? Besides the well-known distortions related to nonsampling errors that affect the quality of information, there is the question of dealing with the influence of external factors. Many monitoring processes simplify causal linkages, particularly those related to farmer participation. For example, an enormous growth in farmers planting fennel without pesticides could be registered two years after the intervention of AS-PTA Nordeste. But this may coincide with a new rural credit scheme for the cultivation of traditional cash crops such as fennel. Isolating the impact of AS-PTA's efforts from the influence of the new subsidy becomes a tricky matter. Farmers' assessment of key influences becomes an essential guide.

AS-PTA ultimately intends for its work to have a regional impact. It is consciously creating alliances and working with institutions and groups that will enhance the scaling up of efforts. Although initially starting with a stakeholder triangle of STR, farmers, and itself, tentative contact is being made with local universities and government research and extension agencies. However, the NGO-government collaboration is not an easy alliance for either party, with political differences between the two institutions causing mutual suspicion throughout Brazil.[15]

Following on EPAGRI's work, AS-PTA sees the municipal councils themselves as another potential route for policy advocacy. Local spending on agricultural services is conducted through the councils, which is potentially a powerful tool to enhance the viability of smallholder agriculture. AS-PTA is seeking ways to influence municipalities and their funds via STR and farmer-led advocacy. But again, political differences between STR and AS-PTA on the one hand, and the local government dominated by larger landowners on the other, will not make collaboration straightforward. Most efforts will, for the time being, be invested in strengthening the stakeholder triangle itself.

This chapter opened with the question: Is the participatory development of agriculture worthwhile, and will it lead to more sustainable, environmentally sound, and socially inclusive forms of agriculture? The evaluation studies of the World Neighbors and EPAGRI experiences give strong indications that, yes, participation works to enhance agricultural development that regenerates natural resources; it can be accomplished without high levels of external inputs, and it can also work for capital-poor smallholders.

Both the EPAGRI and WN studies show that yields can increase, crop diversity can be strengthened, erosion controlled, and soil fertility enhanced. They show that farmer and group capacities for planning, experimentation, and action can be improved and made more self-sufficient. EPAGRI's experience shows that financial investment is worthwhile and that the scope of the work can be extremely large. WN programs show that such efforts can be sustained after program support is withdrawn. Such success is possible when one deals with sustainable agriculture as an interaction of social and biophysical processes.

But the Central American examples also highlight the importance of caution when making claims to success. The moment of measuring impact is central when discussing the sustainability of interventions. If programs are young, external institutions are often still intensely involved in the support of group initiatives, such as in the case of EPAGRI and AS-PTA. Yet the longer-term reality does not always correspond to the rhetoric of those executing projects, as Bunch and López (1995) show. When can one claim that the results obtained, particularly benefits, will endure? Claims to success must be related to short-term, mid-term, and long-term perspectives (Frans Doorman personal communication). Five to ten years after groups or communities have been left to their own devices, with program support withdrawn, often a different picture emerges from the one seen just after program termination. Monitoring over long periods of time is vital if the sustainability of agricultural development is to be understood better and encouraged, particularly when it concerns processes based on social transformation.

This raises a second question: When is such development likely to be worthwhile? Several lessons stand out from the studies:

■ Sustainable agriculture and participatory development are not the sole province of NGOs, and the state can work effectively, as in EPAGRI's example.

- Innovative institutional alliances with municipalities and industries provide important mechanisms for sustaining and spreading positive impacts.

- It is important to encourage the development of strong social identities by supporting the revival or creation of local groups that can continue with agricultural adaptation.

- Information and transitional support—such as the PROSOLO funding, innovative technologies as in WN, or the seed bank initiatives of AS-PTA and the trade unions in Solanea and Remigio—must be provided.

Although the studies contain many rich lessons, one in particular deserves special mention. The emphasis on strengthening local capacities—to prepare PIPs in the case of EPAGRI, to share experimental outcomes in the case of WN, and to monitor collaboration as with AS-PTA—is best left to the words of Bunch and López:

> Make the main goal of programme planning a system whereby farmers learn to, and become motivated to, continue developing their own agriculture. The book "Rural Development, From Farmer Dependency to Farmer Protagonism" (FAO, 1993) expresses the idea well: avoid all dependency. Make sure that each programme role is gradually taken over by the farmers. (1995, 15)

In the light of these answers, a third question looms: What is the value of assessing such work? The earlier answers present obvious value. There is now clear information that certain forms of agricultural development are worth the effort. The second value is an understanding of the possible conditions for achieving and sustaining these benefits. Many of these lessons are not new, although the scale of EPAGRI's work and the postproject termination perspective of the WN programs provide new insights. Many organizations, such as AS-PTA, have built their approach on an analysis of the successes and failures of programs similar to WN and EPAGRI, which have generated guidelines for good practice. EPAGRI itself was built on the merits of older successful initiatives in the neighboring state of Parana, following a critical analysis of its own relatively ineffective earlier approach.

The value for external agencies is clear, but what role does monitoring and evaluation play locally? Who has learned in this process of assessing agricultural development? Have farmers in Santa Catarina, Guatemala, and Honduras learned by being involved in these evaluation studies? In EPAGRI's work, systematic farmer-based M&E remains an obvious challenge. For WN's work, it is unknown how the farmers were involved in the study other than as informants. The studies were undertaken only after encouragement from an external organization—in this case, IIED and the New Horizons initiative. Both are one-shot studies. EPAGRI, COSECHA,[16] and WN might find the evaluation helpful for future work. But M&E has much greater local potential than providing lessons for external agencies. It can also benefit farmers, as long as they are involved in the process beyond simply acting as informants.

In Brazil, AS-PTA is seeking to realize that potential by incorporating M&E processes into its strategy for capacity building and organizational development. First, it views M&E as the prime route to generating information that can feed into higher levels of policy advocacy. Also, farmer-based M&E provides a structured process of reflection for the actors engaged in agricultural development. Reflecting on achievements helps motivate the partners, and assessing limitations allows for the adjustment of plans. Both motivation and modification are essential for sustainability of impacts.

M&E is also a process of mediating between the actors. Starting small— with STR, farmers, and itself—AS-PTA is using the negotiations around an M&E process to forge new and stronger alliances. Negotiations revolve around the clarification of objectives in the joint venture to enhance small-holder production, around defining success in the collaborative efforts and the relative importance of social processes versus biophysical measurements, and around defining roles and responsibilities. Bringing together farmers who are not organized in terms of their agricultural activities to test, monitor, and evaluate agricultural innovations provides a mechanism to create new group-based social identities. This focus on the importance of strong local groups and alliances between them highlights AS-PTA's serious attempts to avoid dependency on itself as a temporary external agency. Development of the M&E process with STR and farmers is creating local capacity to build other monitoring processes, after AS-PTA moves on to work in other municipalities.

These negotiations take time. Testing an M&E system and ensuring that it is viable and useful will take three years in Paraiba.[17] It is not as simple as assigning a set of indicators and a methodology for recording, analyzing, and sharing. Undertaking processes based on high levels of participation means investing energy in building local capacity rather than doing the work for others. Success will be slow, but scaling up and diffusion can become a sustained process if farmers take on all roles, including those of monitoring and evaluation.

However, participatory development of agriculture will not necessarily provide all the answers.[18] It is a slow, highly political process (whether organizations care to admit it or not), one in which mediation between diverse objectives and agendas is central. This requires great organizational flexibility, sensitivity, and patience. It means a focus on working to strengthen people's capacities to reflect and act independently of external support. Because capacity is generally considered an intangible good and is not easily quantifiable, this presents new challenges for monitoring and evaluation. As the mechanism for achieving sustainable agriculture shifts from technologies to capacities, from fields to watershed, so the focus of evaluation shifts.

Learning from the past is essential to working better in the future. But "[a] major shortcoming of the present renewed discourse on participation is its disregard of even recent history, the conceptions and prescriptions that flourished a few years ago . . . the [inherent] risk of 'inventing gunpowder in the twentieth century'" (Stiefel and Wolfe 1994, 3). Many organizations and agencies will continue to "invent gunpowder," as they find it too painful to look back in order to think forward. Organizational reflection and learning with rural people are often inhibited by paralyzing internal structures, inadequate staff incentives, overspecialization, flawed field methodologies, and crippling internal financial procedures (notably the speed of disbursements).[19] The learning imperative applies not only to method but also to the institutional forms that allow learning. Monitoring and evaluation with all stakeholders, particularly of participatory programs, offer vital opportunities for motivation and modification of existing efforts in sustainable agriculture.

Notes

1. Thanks to Gaston Remmers (1994) for this metaphor.
2. Calculated at 1,266 worker days per hectare; US$1.50 daily rate, or a total cost of US$1,900 per hectare.
3. An annotated bibliography on cases of participatory watershed development and participatory rural appraisal contains over 100 examples (IDS 1995), and there are many more farmer-centered initiatives that have not labeled themselves as PRA.
4. Hudson (1991) urges that evaluations be based on the extent to which a program has achieved its own objectives. But considering the difficulty of defining objectives in terms of sustainability and the lack of clear objectives in general, this approach presents serious limitations.
5. A word of caution is required regarding PME and PIM, which are not necessarily as participatory as their names imply. A critical case-by-case review is necessary to determine how participation is interpreted.
6. More experiences in this area can be found in the health sector (compare Feuerstein 1986; Narayan 1993; Koning and Martin 1996).
7. Several typologies of participation exist that try to cluster the practice of participation into common interpretations (see Cornwall 1996; Stiefel and Wolfe 1994; Biggs 1989, Adnan et al. 1992; Arnstein 1969; Hart 1992; Pretty 1995). Such typologies, however, hide several pitfalls that make their use in evaluation processes problematic; there is no clear linear path from coercion to participation and from exclusion to inclusion (see Guijt 1991; Stiefel and Wolfe 1994; MacGuire 1996).
8. The EPAGRI section draws extensively on the study by Valdemar Hercilio de Freitas (1995), commissioned as part of an IIED collaborative research program called New Horizons: The Economic, Social and Environmental Impacts of

Participatory Watershed Development. Sincere acknowledgments to the author for permitting the use of raw data from his study. The analysis of the material in this context is my own, as is any error.

9. Over 100 experiments are being conducted by EPAGRI's research branch to search for solutions to problems identified in the participatory watershed development process.

10. A recent agreement with the World Bank has handed responsibility to the State Planning Institute, CEPA, for a three-stage microcatchment-level evaluation: before PNMH, in the middle, and after PNMH stops.

11. This section draws directly on Bunch and López 1995. The authors' permission to cite directly is thankfully acknowledged. This study was also part of the IIED's collaborative research program New Horizons: The Economic, Social and Environmental Impacts of Participatory Watershed Development. For a summary of the findings of the New Horizons research program, see Hinchcliffe et al. 1995.

12. No other programs have worked for any significant period of time in these villages on soil conservation or crop production since program termination, except to some extent in Pacayas and Pacoj, and yields in nearby nonprogram villages are low, averaging less than 1.6 tons per hectare.

13. AS-PTA was instrumental in the establishment and development of PTA, a national network of about twenty-three NGOs working with similar issues.

14. This is the basis of a three-year collaboration with IIED, funded by the Overseas Development Agency, U.K.

15. EPAGRI, for example, has virtually no contact with local NGOs, although their working approaches are similar.

16. COSECHA is a Central American advisory NGO that undertook the evaluation of its previous work as farmer extensionists in soil conservation projects in Honduras and Guatemala.

17. A parallel process is taking place in Minas Gerais with another NGO, CTA-ZM, to allow for a comparison of the content of the monitoring methodology and of the process of developing it.

18. The difficulty of this political process is apparent in the poor quality and limited forms of much of what passes as "participatory" development elsewhere (Guijt and Cornwall 1995).

19. See also Korten 1980; Chambers 1992; Pretty 1995; Thompson 1995; Hudson 1992.

References

Adnan, S., A. Barrett, S. M. Nurul Alam, and A. Brustinow. 1992. *People's Participation: NGOs and the Flood Action Plan*. Dhaka, Bangladesh: Research and Advisory Services.

Andrade, M. C. 1980. *A Terra e o Homen no Nordeste*. São Paolo: Hucitec.

Arnstein, S. R. 1969. A Ladder of Citizens' Participation. *Journal of the American Institute of Planners* 35 (July): 216–24.

Bentley, J., and K. Andrews. 1996. *Through the Roadblocks: IPM and Central American Smallholders*. Gatekeeper Series no. 56. London: IIED.

Biggs, S. 1989. *Resource-Poor Farmer Participation in Research: A Synthesis of Experience from Nine National Agricultural Research Systems*. OFCOR Project Study no. 3. The Hague: ISNAR.

Bishop, J. 1992. *Economic Analysis of Soil Degradation.* LEEC Gatekeeper Series no. GK 92–01. London: IIED.

Bunch, R. 1977. Better Use of Land in the Highlands of Guatemala. *In* Growing Out of Poverty, *edited by E. Stamp. Oxford: Oxford University Press.*

Bunch, R. 1988. Guinope Integrated Development Programme, Honduras. In *The Greening of Aid: Sustainable Livelihoods in Practice,* edited by C. Conroy and M. Litvinoff. London: Earthscan Press.

Bunch, R. 1990. *Low Input Soil Restoration in Honduras: The Cantarranas Farmer-to-Farmer Extension Programme.* Gatekeeper Series no. 23. London: IIED.

Bunch, R., and G. López. 1995. *Soil Recuperation in Central America: Sustaining Innovation After Intervention.* Gatekeeper Series no. 55. London: IIED.

Busscher, W. J., D. W. Reeves, R. A. Kochhann, P. J. Bauer, G. L. Mullins, W. M. Clapham, W. D. Kemper, and P. R. Galerani. 1996. Conservation Farming in Southern Brazil: Using Cover Crops to Decrease Erosion and Increase Infiltration. *Journal of Soil and Water Conservation* 51 (3): 188–92.

Campbell, A. 1994. Participatory Inquiry: Beyond Research and Extension in the Sustainability Era. Paper prepared for the International Symposium on Systems-Oriented Research in Agriculture and Rural Development, Montpellier, France, November.

Chambers, R. 1992. *Challenging the Professions: Frontiers for Rural Development.* London: IT Publications.

Cornwall, A. 1996. Towards Participatory Practice: Participatory Rural Appraisal (PRA) and the Participatory Process. In *Participatory Research in Health: Issues and Experiences,* edited by K. de Koning and M. Martin. London: Zed Books.

Cornwall, A., I. Guijt, and A. Welbourn. 1993. *Acknowledging Process: Challenges for Agricultural Research and Extension Methodology.* Discussion Paper no. 333. Brighton: IDS.

Critchley, W. R. S., C. Reij, and T. J. Willcocks. 1994. Indigenous Soil and Water Conservation: A Review of the State of Knowledge and Prospects for Building on Traditions. *Land Degradation and Rehabilitation* 5 (4): 293–314.

Davies, R. 1996. An Evolutionary Approach to Facilitating Organisational Learning. An Experiment by the Christian Commission for Development in Bangladesh, Mimeo.

de Graaff, J. 1993. *Soil Conservation and Sustainable Land Use: An Economic Approach.* Amsterdam: KIT.

Feuerstein, M. T. 1986. *Partners in Evaluation: Evaluating Development and Community Programmes with Participants.* London: TALC/Macmillan Press.

Freitas, V. H. de. 1995. Transformations in the Micro-catchments of Ribeirão das Pedras/Agrolândia and of Rio Macacos / São Lourenço do Oeste, Santa Catarina—Brazil. Paper presented at New Horizons: The Economic, Social and Environmental Impacts of Participatory Watershed Development, IIED international workshop, Bangalore, India, December.

Gardner, Gary. 1996. Preserving Agricultural Resources. In *The State of the World 1996.* London: Earthscan.

Guijt, I. 1991. *Perspectives on Participation: An Inventory of Institutions in Africa.* London: ED/FTP.

Guijt, I. 1994. Trip Report to Brazil, July 25 to August 24, 1994. IIED internal memo.

Guijt, I. 1996. Participatory Monitoring in Sustainable Agriculture: An Introduction to the Key Elements. IIED unpublished report.

Guijt, I., and A. Cornwall. 1995. Critical Reflections on the Practice of PRA. *PLA Notes* 24: 2–7.

Guijt, I., and P. Sidersky. 1996. Agreeing on Indicators. *ILEIA Newsletter* 12 (3): 9–11.

Hart, R. 1992. *Children's Participation: From Tokenism to Citizenship*. Florence: UNICEF/International Child Development Centre.

Hinchcliffe, F., I. Guijt, J. N. Pretty, and P. Shah. 1995. *New Horizons: The Economic, Social and Environmental Impacts of Participatory Watershed Development*. Gatekeeper Series no. 50. London: IIED.

Hudson, N. 1991. *A Study of the Reasons for Success or Failure of Soil Conservation Projects*. Rome: FAO.

Hudson, N. 1992. *Land Husbandry*. London: B. T. Batsford.

IDB. 1995. *Concepts and Issues in Watershed Management*. Working Paper no. 2–95. Washington, D.C.: Inter-American Development Bank.

IDS. 1995. Participatory Rural Appraisal: Abstracts of Selected Sources, Institute of Development Studies, Brighton.

IIED. 1992. New Horizons. The Social, Environmental and Economic Impact of Participatory Watershed Development. IIED research proposal.

Koning, K. de, and M. Martin, eds. 1996. *Participatory Research in Health: Issues and Experiences*. London: Zed Books.

Korten, D. 1980. Community Organisation and Rural Development: A Learning Process Approach. *Public Administration Review* 40 (5): 480–511.

Lutz, E., S. Pagiola, and C. Reiche. 1994. The Costs and Benefits of Soil Conservation: The Farmers' Viewpoint. *World Bank Research Observer* 19 (2): 273–95.

MacGuire, P. 1996. Proposing a More Feminist Participatory Research: Knowing and Being Embraced Openly. In *Participatory Research in Health: Issues and Experiences,* edited by K. de Koning and M. Martin. London: Zed Books.

Magrath, W. B., and J. B. Doolette. 1990. Strategic Issues in Watershed Development. In *Watershed Development in Asia: Strategies and Technologies*, edited by J. B. Doolette, and W. B. Magrath. Technical Paper no. 127. Washington, D.C.: World Bank.

Martin, P. 1991. Environmental Care in Agricultural Catchments: Toward the Communicative Catchment. *Environmental Management* 15 (6): 773–83.

Merrill-Sands, D., S. Biggs, R. J. Bingen, P. Ewell, J. L. McAllister, and S. Poats. 1991. Institutional Considerations in Strengthening On-Farm Client-Oriented Research in National Agricultural Research Systems: Lessons from a Nine-country Study. *Experimental Agriculture* 27 (4): 343–73.

Narayan, D. 1993. *Participatory Evaluation Tools for Managing Change in Water and Sanitation*. World Bank Technical Papers. Washington, D.C.: World Bank.

NRC (National Research Council). 1989. *Alternative Agriculture*. Washington, D.C.: National Academy Press.

Oldeman, L. R., et al. 1991. *World Map of the Status of Human-Induced Soil Degradation: An Explanatory Note*. 2d ed. Nairobi and Wageningen, Netherlands: International Soil Reference and Information Centre, UNEP.

Ostberg, W., and C. Christiannson. 1993. *Of Lands and People*. Working Paper no. 25. Stockholm: Environment and Development Studies Unit, Stockholm University.

Paul, J. M. 1994. Avaliaçao Qualitativa do PROSOLO/Microbacias. Resumo das Principais Conclusoes e Recomendacoes. Mimeo. Instituto CEPA/SC.

Pla-Sentis, I. P. 1992. Soil Conservation Constraints on Sustained Agricultural

Productivity in Tropical Latin America. In *Soil Conservation for Survival*, edited by K. Tato and H. Hurni. Ankeny, Iowa: Soil and Water Conservation Society.

Pretty, J. N. 1995. *Regenerating Agriculture: Policies and Practice for Sustainability and Self-Reliance*. London: Earthscan.

Pretty, J. N., and P. Shah. 1994. *Soil and Water Conservation in the Twentieth Century: A History of Coercion and Control*. Rural History Centre Research Series no. 1. University of Reading.

Pretty, J. N., J. Thompson, and J. K. Kiara. 1995. Agricultural Regeneration in Kenya: The Catchment Approach to Soil and Water Conservation. *Ambio* 24 (1): 7–15.

Reij, C., S. D. Turner, and T. Kuhlmann. 1986. Soil and Water Conservation in Sub-Saharan Africa: Issues and Options. Rome: IFAD.

Reijntjes, C. C., B. Haverkort, and A. Waters-Bayer. 1992. *Farming for the Future: An Introduction to Low-External-Input and Sustainable Agriculture*. London: ILEIA/Macmillan Press.

Remmers, G. 1994. Endogenous Development in Traditional Rural Areas: Hitting a Moving Target. Paper presented at the International Seminar, "Towards Regional Plans for Endogenous Rural Development in Europe," ISEC, Cordoba, Spain, December 1994.

Rugh, J. 1986. *Self-Evaluation: Ideas for Participatory Evaluation of Rural Community Development Projects*. Oklahoma City: World Neighbors.

Shah, P., and M. K. Shah. 1994. Impact of Local Institutions and Para-professionals on Watersheds: Case Study of AKRSP in India. Paper presented at New Horizons: The Economic, Social and Environmental Impacts of Participatory Watershed Development, IIED international workshop, Bangalore, India, December.

Sidersky, P. 1995. Desenvolvimento Local, Pequeños Agricultores e Participaçao. AS-PTA. Manuscript.

Sommer, M. 1993. Whose Values Matter? Experiences and Lessons from the Self-Evaluation in PIDOW Project. Bangalore, India: Swiss Development Corporation.

Stiefel, M., and M. Wolfe. 1994. *A Voice for the Excluded. Popular Participation in Development: Utopia or Necessity?* Geneva and London: UNRISD/Zed Books.

Thompson, J. 1995. Participatory Approaches in Government Bureaucracies: Facilitating the Process of Institutional Change. *World Development* 23 (9): 1521–54.

Treacy, T. J. M. 1989. Agricultural Terraces in Peru's Colca Valley: Promises and Problems of an Ancient Technology. In *Fragile Lands of Latin America: Strategies for Sustainable Development*, edited by J. O. Browder. Boulder, Colo.: Westview Press.

Uphoff, N. 1992. *Learning from Gal Oya: Possibilities for Participatory Development and Post-Newtonian Science*. Ithaca, N.Y.: Cornell University Press.

WRI (World Resources Institute/UNEP/UNDP/World Bank). 1996. *World Resources 1996–1997*. Oxford: Oxford University Press.

Part III

Markets as Mediation

5

Moving Beyond Banana Trade Wars, 1993–96

Mediation in Solidarity for Sustainability

Alistair Smith

> International trade should be conducted with the objective of improving
> the well-being of people, whilst recognising the need to promote socially
> just and ecologically sustainable development and prudent resource
> management, in accordance with the precautionary principle, trans-
> parency and participatory democracy. ("Alternative Treaty" 1992)

The promotion of sustainable banana production and trade in cooperation
with plantation workers, family farmers, and other European nongovernmental organizations (NGOs) is the central long-term aim of the new
Norfolk, England-based nonprofit cooperative Banana Link (BL). Although
formally registered in January 1996, BL had evolved over three preceding
years into a complex process of coordinating a dynamic information network and facilitating the building of an international alliance around the
objective of sustainability in the banana sector.

Behind these noble yet ill-defined aims and objectives, however, lie profound economic and democratic challenges for those of us now engaged in
this process of change. There has historically been a political minefield sepa-
rating producers and consumers, both from each other and from their counterparts in other countries or regions. It is across this minefield that we at BL
have tried to pick a route. It is those in the middle, the corporate "economic
intermediaries" between the producers and consumers, who have provoked
the banana trade wars. It is these companies that have manipulated the
international institutional framework to their best advantage. In particular,
the biggest banana company in the world, Chiquita Brands International,
has been successful in persuading the U.S. Congress and the White House

---

This chapter is written by a mediator from a personal perspective. The style maintained throughout the text, therefore, is consciously that of a personal account of
events and of the author's personal analysis.

that its corporate interests are also those of the U.S. government and its major Latin American governmental allies.

On 9 May 1996, in Geneva, the formal establishment of a dispute "panel" in the World Trade Organization (WTO) was announced: between the fifteen European countries on the one hand, and the United States, Guatemala, Ecuador, Honduras, and Mexico on the other. A panel of three selected lawyers is to arbitrate on the vexed question of whether the European Union's single-market banana regime (known as "404/93") is discriminatory to U.S.-based corporate and Latin American governmental interests. This legal challenge followed two earlier dispute panels of the General Agreement on Tariffs and Trade (GATT) regarding the banana trade policy of the European Union (EU). This challenge was also preceded by well over a year of threatened sanctions against the EU as well as Costa Rica and Colombia (using punitive U.S. domestic trade law section 301) and by numerous cases—both resolved and unresolved—brought by German governmental and corporate interests before the European Court of Justice in Luxembourg since 1993.

At stake is the question of sovereign rights to "protect" preferred producers with different tariff levels and supply quotas—that is, sovereign rights to determine a country's own national or regional agricultural trade policies. For the longer term, there also emerges the critical question of whether international trade rules will in practice allow countries (or regional trading blocs like the EU-15) to discriminate at their borders between "identical products on the basis of the way in which they were produced or processed" (Article III of GATT), for example, on the basis of being more or less sustainably produced and processed. The whole issue of fair trade criteria for bananas and the social and environmental aspects of trade is now directly affected by the esoterically worded judgments made in such transparent and inaccessible forums as the new WTO. Impenetrable institutional language and unaccountable institutional practices pose a much wider problem of democracy at the intergovernmental level. Banana trade disputes would appear to herald major conflicts over the future rules of international exchange.

The reality today, however, is that the plantation workers' unions, the family farmers' organizations, and a broad diversity of other NGOs currently active in this new alliance are all committed to changing the terms of the banana trade for good. Furthermore, there is now a widely shared understanding within the alliance that we will all continue to collaborate at the intercontinental level for at least the next five or six years and that a fair trade niche in the European banana market has been no more than the initial unifying objective for the European NGOs at this end of the alliance. What we are in fact working on is more akin to a new international economic order in the banana sector.

So what is special about bananas? Why has this informal alliance emerged, spanning producers and consumers in over thirty countries? Who are the main actors interested and active in trying to move beyond the current macroeconomic and geopolitical wars taking place over bananas? What processes have been evolving since the intervention of Farmers' Link (FL), a small NGO based in East Anglia in the United Kingdom working on the promotion of a transition to sustainable agriculture and rural development (SARD) worldwide? What, if anything, has changed as a result of mediation in this complex, fast-moving, and highly conflictive sector of the world's agrifood economy by a small NGO motivated by "solidarity"? What do these experiences tell us about what works and what does not, and hence what actions may be effective in the future? These are the key questions that I—as informal (self-appointed), then formal (collectively mandated) coordinator of this process since 1992—have asked in recording the following highly subjective version of events.

Why Pick Bananas?

Bananas are the fifth most important agricultural commodity in world trade after cereals, sugar, coffee, and cocoa. For at least fifteen Latin America and Caribbean producer countries, the Cavendish banana is a crucial source of export income.[1] In 1995, world banana trade was valued at over US$8 billion, with the EU being the largest importer, consuming nearly 40 percent of traded bananas. Each of the 350 million EU citizens now consumes an average of just over 10 kilograms of bananas per year.

Apart from the 10 percent of bananas and plantains produced annually that are traded on the world market, the crop is grown by millions of small-scale farmers in Africa, South Asia, and northern Latin America for household consumption and/or local markets. Bananas are the fourth most important staple food worldwide, making a significant contribution to food security in dozens of countries in the tropics. Most of this production is achieved with few or no external inputs. However, once a producer grows for export markets to consumers in the industrialized world, considerable and growing levels of external inputs (seed, chemicals, fertilizer) are required to effectively compete in those markets. The range of negative economic, social, and ecological impacts associated with production for export in all producer countries is increasingly well known in the consumer countries.

The EU's banana trade is dominated by a classic oligopoly of six companies that controlled at least 85 percent of world trade in 1994. Indeed, the pressures in the industry meant that by 1996, only four of those big-six multinational companies remained in the banana business. Historically, these companies have been able to maintain high levels of profitability from their banana production and trading activities. The very low returns to family farmers (5 to 12 percent) and plantation workers (1 to 3 percent),[2] as a proportion of the consumer price, mean that this fruit has become a classic example of inequitable primary commodity trade. In May 1995, for example, the Association of Panamanian Banana Growers reported that the price to producers had fallen from $5.25 per 18.4-kilogram case in 1992 to $3.86. Meanwhile, a European wholesaler sells Panamanian bananas to retailers at around $26.[3] Plantation workers and smaller-scale farmers have been the first to suffer from the generalized fall in producer prices since 1992.

In summary, the current patterns of production, exchange, and consumption make bananas a strong pedagogical example of how the interests of both producers and consumers are marginalized in an increasingly unsustainable world commodity trade. Many of the most difficult issues at the heart of the search for transitional mechanisms favoring greater sustainability can be drawn out by reference to the case of the European banana trade.

The case for concerted action by European NGOs whose agenda centers on a transition to sustainable development is therefore compelling. Bananas are a product that interests consumers, more so than soya or wheat, for example. Additionally, NGO campaigns and initiatives in the 1970s in Switzerland, and in the 1980s and early 1990s in several European Community countries, had already raised some of the issues around which the mediation of FL—and now BL—has been concentrated. The different elements of concern about the economic, social, and environmental conditions of banana production were brought out, without specifically using the concept of sustainability to encapsulate an alternative to all these trends.

But the real trigger that precipitated an international response at all levels can be singled out as having been the 1992 proposal and then ratification of an EU market regulation for the Common Organization of the European Community Market in Bananas,[4] which was implemented in July 1993. This EU regulation, the most complex and controversial in the history of the European Community edifice, triggered the whole range of actors described in this chapter to take up their own strongly polarized positions. At the heart of this regulatory agreement is an EU import policy covering the next ten years that places quantitative restrictions on bananas from Latin American plantations, which preserves free access to the EU market for bananas produced in a dozen African and Caribbean ex-colonies of the United Kingdom, France, the Netherlands, Italy, Spain, and Portugal and which offers direct income support to "domestic" suppliers in the Canary Islands, French West Indies, Madeira, and Greece. At the time, it was clear that this brave but ultimately technocratic attempt to "square the circle" of the EU's GATT and Lomé commitments was going to satisfy either everybody or nobody. In the end, almost nobody was satisfied, and some have gone to war over it or because of it.

Finally, in answering the question of why bananas, it should be highlighted that the innocent banana has been, since the last minutes of the signing of the Treaty of Rome nearly four decades ago, the most sensitive and politicized of subjects in intergovernmental policymaking in Europe. Postwar Germany was given a "concession" by the U.S. government and its big banana companies, which in turn became enshrined in European Community trade policy: no tariff was to be charged on entry of Latin American plantation bananas to West Germany, whereas imports into the other five member states attracted a 20 percent tariff. This exception in the founding treaty of the European Community of six, known as the Adenauer Clause, survived until the Single European Market of 1993.

Then, with the signing of the Lomé Convention in 1975, the United Kingdom, France, the Netherlands, and Italy succeeded in enshrining their own preferential arrangements for banana imports (also no tariff on entry) with their colonial or ex-colonial suppliers. So, by 1996, France, Spain, and, to a lesser extent, the United Kingdom, as chief architects and defenders of the new single EU banana regime, were pitted against Germany, Denmark, and the Benelux countries. In practice, this rift between member states, expressed in the decision-making body of the Council of Agriculture Ministers, is paralyzing any attempt to amend the 404/93 regulation. Indeed, any issue that polarizes the two key architects of the European Community as much as banana policy does will always pose a threat to the very foundations of intergovernmental institutions in western Europe. This simple geopolitical fact guarantees that interventions in the banana sector will continue to generate political heat.

Other Triggers in the Process

The original request for FL to become involved in the dissemination of information on banana trade policy came in the summer of 1992, from the program officer of the Windward Islands Farmers' Association (WINFA), a confederation of small farmers' organizations from the four English-speaking Windward Islands and Martinique. WINFA was realizing that if small-scale banana farmers wanted to have an independent voice in Europe apart from their own governmental lobby and the Banana Growers' Associations (BGAs), which the governments controlled, they would need to build up their own sources of information. In order to generate their own positions autonomously of government, the four island BGAs, and their British NGO funders (Oxfam and Christian Aid), WINFA asked FL to start playing a role in accessing and distributing information about the forthcoming Single European Market regime. Later that year, WINFA asked FL to organize a European lobbying tour for the WINFA program officer around the crucial EU Heads of State Summit hosted by the United Kingdom in Edinburgh and to identify other European NGOs susceptible to offering either political or practical solidarity. So FL's initial involvement stemmed from a direct request in St. Lucia to play specifically defined informational, logistical, and networking roles in support of WINFA's attempts to build its own institutional capacity.

From the outset, discussions were also initiated between FL and WINFA on how to access more information about conditions in the Latin American plantations, which small Caribbean farmers were increasingly going to have to compete with. Were there small farmer exporters in Latin America too? And what was their monopoly exporting and marketing company doing establishing plantations in Costa Rica? These were the kinds of unanswered questions that led in 1993 to the first joint delegation of Latin American plantation workers and eastern Caribbean small farmers. Although a workers' representative came to Germany and Belgium in March 1993, at this stage, the critical contact between German NGOs and British NGOs had not yet been made. It is telling of the situation at the time that trade union leader Gilberto Bermudez, on his return home, was denounced as a "bad Costa Rican" by his own government.[5]

The joint visit to Europe in October 1993, facilitated by FL along with Belgian and German NGO colleagues, was the first outward signal to the companies, and to the governmental and intergovernmental actors, that European NGOs, even small and unfamiliar ones, could act as a powerful catalyst. They could project and even strengthen the common interests of plantation workers and small farmers, notably across traditional divides within Europe, at a time when governments had become firmly aligned in two camps for and against the new EU banana regime—Latin American versus African, Caribbean, and Pacific (ACP).

It was the first time that British NGOs, British government, and the British public had heard from Latin American plantation workers firsthand. Similarly, in Germany, NGOs, government, and the public heard for the first time from small-scale farmers in the Caribbean. This was to be the start of a process of transformation of wider awareness in Europe: from a situation in which consumers tended to know only about their own preferential suppliers, to one in which a more complete picture of the conditions of production and trade in all the EU's suppliers could be presented, analyzed, and acted on. The perspective of long-term sustainability of economic, social, and ecological conditions of production, which informed both FL and a few other European NGO actors at this time, allowed us to develop our analysis and critique of an industry that was increasingly externalizing its social and ecological costs as banana prices fell and a new process of market concentration started to take place.

Who Are the Actors, and What Do They Say and Do?

Workers, especially those engaged in production for export have played a determining role in the modern history of Latin American societies. Their struggle for material well-being and control over their own lives has fundamentally altered the direction of national political evolution and the pattern of economic development of the countries of the region. (Bergquist 1986, 10)

The major actors on the scene can be categorized in five groups.

Unions

This group consists of independent plantation workers' unions of employees on medium and large-scale plantations in Latin America and the Philippines, and other trade unions in the agrifood sector worldwide. Our key partner in the alliance is the Coordinadora de Sindicatos Bananeros de America Latina, based in Costa Rica and Honduras.

By 1992, all Latin American plantation workers' trade unions found themselves seriously weakened, as compared with a decade earlier. Many benefits acquired over decades of struggle had been eroded or were under threat. In some cases, unions had gone bankrupt. *Solidarismo*, a kind of management-friendly workers' association, had been imposed by most banana companies as an alternative to the "confrontational" style of the independent trade unions. The sheer capital and political will of the companies and governments behind this so-called *solidarismo* movement, coupled with its superficial material perks in the social and cultural spheres, made it difficult for many trade unions to survive.

But this was one of the key factors in propelling a process of concertation for survival, which manifested itself in the emergence of the Coordinadora de Sindicatos Bananeros de Centroamerica y Colombia in 1993. Some twenty-five unions representing tens of thousands of plantation workers in seven countries were starting to speak with one voice for the first time. At a hearing of the European Parliament in March 1993, a Costa Rican union leader from the Coordinadora surprised many by declaring support for the EU's proposed tariff quota on banana imports from Latin America, on the grounds that a stable, guaranteed market would at least allow plantation workers the political space to negotiate better conditions within their own countries.

Two other regional conferences of banana workers in 1994 and 1995, plus a regular program of strategy meetings and regional training seminars on key issues facing banana workers, have brought this process of unification in Central America and Colombia to early maturity. Parallel and related to this process, national banana union coordination offices have emerged in both Costa Rica and Honduras; similar processes are being discussed in Guatemala and Panama, potentially overcoming decades of cold war ideological polarization within the union movement. Additionally, the coordinating bodies of regional and national unions are now actively extending their contacts with banana unions in the Philippines, Belize, Suriname, and Venezuela, as well as with food and agricultural workers' unions in Europe.

Although previously only weakly linked to any international trade union structures, the banana workers have started to articulate their regionally agreed concerns via the International Union of Food-workers. This union itself only recently merged with the International Federation of Plantation, Agricultural and Allied Workers, to form a confederation of 320 food sector trade unions in over 80 countries. The multiple crises facing the banana workers' unions since the mid-80s has, however, led them to a redefinition of the role of independent workers' organization: *un nuevo sindicalismo* (a new trade unionism). Although very much a product of internal debates, this process of redefinition can also be partially attributed to the impact of the new solidarity links which we have forged within the international alliance.

Farmers' Organizations

The second group consists of small-scale farmers' organizations in the Caribbean and Ecuador (members have landholdings of 0 to 50 acres) producing bananas for the European market and other peasant farmer unions in Latin America, Asia, and Africa. Our key partner is WINFA, based in St. Vincent.

Currently, there are virtually no small-scale banana exporters in Latin America. Only in Ecuador, Honduras, and the Dominican Republic are there a few thousand. The vast majority are in the Windward Islands—some 25,000. There are also about 10,000 in the Canaries, 500 in Jamaica, and a few hundred more in the French islands of Martinique and Guadeloupe. Taken together, these 40,000 or so small exporters currently supply less than 15 percent of the EU market, with a maximum potential to supply about 20 percent. In Ecuador, still the largest exporter in the world, the small farmer exporters are used as a kind of buffer stock, getting squeezed out of the market as soon as the world market shrinks slightly. This happened in early 1993 as the market became saturated and producer prices fell sharply. In the Windward Islands, a combination of world price, sterling devaluation, drought, and four major hurricanes in 1994 and 1995 has conspired to squeeze the most economically fragile farmers out of the banana market.

Both WINFA in the eastern Caribbean and UROCAL, an organization of small-scale banana producers in Ecuador, see diversification within the banana sector (organic or ecological bananas, processing for niche markets and so forth) and beyond it (other cash or food crops or livestock), as well as fair trade, as central to their long-term livelihood strategies. With the acqui-sition of the Geest banana business by a joint venture of WIBDECO (a semi-state-owned company registered in St. Lucia) and Fyffes, banana farmers in the Windwards have the opportunity to secure themselves a fair price—one that covers real costs of production plus a reasonable profit margin.

WINFA is currently in the process of negotiating with the EU and with its own governments and company to play a role not only as initiator and facil-itator of a major diversification thrust among banana producers but also as a lead agency in a future environmental improvement program with small farmers, which would give Windward Island exporters a comparative advan-tage in the increasingly sustainability-conscious European market. However, this flies in the face of the economic wisdom emanating from the U.K.-based consultants Cargill Technical Services (CTS), for which higher external-input use and a clear-out of the smallest-scale banana farmers seems to be a gospel not to be challenged. The ensuing economic debate—both within the island farming communities and between farmers and other actors—over market survival, quality standards, sustainability, and fair trade has been coming into focus since the WIBDECO-Fyffes joint venture broke the foreign export monopoly on bananas. The outcomes of this debate will be critical for the future credibility and success of the wider alliance with plantation workers and critical European consumers.

ASOCODE, a new confederation of Central American small and medium-scale farmers' organizations representing over 600,000 producers in fifteen organizations in seven countries, although not representing banana farmers, has also emerged as an ally in the quest for sustainable banana production, democratic land reform, cooperative marketing, and fair trade. Solidarity has also been actively voiced by the European Farmers' Coordination, repre-senting several hundred thousand smaller-scale farmers in thirteen European countries, and by the international movement, which brings together all these peasant farmer organizations under the umbrella of the Via Campesina (see Chapter 8 in this book).

NGOs

The term NGO covers a wide range of actors—many in Europe, some in the producer countries—working regularly on banana trade issues: development agencies such as Oxfam Belgium, Christian Aid (UK), or Solidaridad (Netherlands); environmental organizations such as Pro Regenwald in Germany or AECO in Costa Rica; solidarity organizations, human rights organizations, and churches at both ends; and alternative trade organiza-tions such as the Swiss-based Gebana or fair-trade labeling organizations such as the Max Havelaar Foundation, Transfair, the Fair Trade Foundation, and FL itself.

Despite the many NGOs working on banana issues, direct cooperation between NGOs with access to different information and therefore different analyses of the banana wars was far from automatic. Indeed, at one stage in late 1993, the Caribbean and Latin American sections of some northern European development agencies were at loggerheads over banana issues. NGOs in Germany, Denmark, and the Benelux countries tended to lack information on and therefore understanding of the English-speaking Caribbean producers, whereas the British, French, and Spanish NGOs knew about production only in their ex-colonies or current overseas territories.

Beyond organizational issues was a resourcing constraint. By the mid-1990s, many of the big NGOs had cut or severely reduced their programs in many of the banana exporting countries of Latin America and the Caribbean. Their rationale was that poverty was more acute and widespread in other regions of the South, despite the continued existence of real pockets of poverty and ill health in communities that lived off a cash crop that was in oversupply and where wages and working conditions were getting worse rather than better.

The European Banana Action Network (EUROBAN) was born both to address this issue of partial information flow and to examine the potential for an informal alliance of European NGOs seeking real solidarity action with the most vulnerable and exploited actors in the banana chain. The coordinator has always sought to set the short-term context of pan-European NGO collaboration over market access for fair-trade bananas within the wider optic of long-term sustainability—of both production and the trade itself. An intensive period of information exchange, translation, and processing took place during 1994 and 1995, with the objective of rapidly overcoming these information gaps and biases. From early 1995, EUROBAN has met quarterly for two-day strategy meetings, always with the participation of one or more representatives of our key partner organizations from the wider alliance. EUROBAN itself remains an informal structure whose coordination tasks are entirely financed by its own participating NGOs, of which there are now thirty-two from thirteen countries.

It was in this context that FL's approach on the matter of bananas developed. This chapter's epigraph summarizes well the approach that underpinned the call for a first meeting of European NGOs interested in the concept of a pan-European fair-trade initiative for bananas in February 1994. If we could make concerted progress on a more sustainable banana production and trade, it would surely facilitate future actions of the same orientation in other sectors. My firm belief as initiator was that our efforts had to be, on the one hand, pan-European (EU and preferably wider) and, on the other hand, undertaken in direct collaboration (to the extent possible) with both plantation workers and small-scale farmers.

Table 5.1   International and National Banana Companies Most Active in the EU Banana Market

Multinational Headquarters	National
Chiquita (USA)	Noboa (Ecuador)
Dole (USA)	Uniban, Banacol (Colombia)
Fyffes (Ireland)	Corbana (Costa Rica)
	JAMCO/Jamaican Producers (Jamaica, UK)
Del Monte (USA, Chile)[1]	WIBDECO (Windward Islands/UK)
Geest (England)[2]	Bananic (Nicaragua, Belgium)[3]
Pomona (France)	Somalfruit (Italy, Somalia)[4]

Notes:
1. Del Monte was bought in mid-1996 by IAT, based in Santiago, Chile.
2. Geest's banana division was sold to a fifty-fifty joint venture of Fyffes and WIBDECO in December 1995. Its eleven Costa Rican plantations were resold by Fyffes/WIBDECO to a consortium of Central American businessmen in March 1996.
3. Bananic has not been selling bananas in the EU market since November 1993. The company currently survives by just trading licenses and leasing quota.
4. Somalfruit's operations have been seriously hampered by continuing civil war since 1991, but it managed to export nearly 20,000 tons in 1995.

The collective response formally agreed on by EUROBAN NGOs in April 1996 was to present the medium-term plans of the three regional nodes of the international alliance as a common work program, and to seek support from the bigger development NGOs in Europe for a coordinated multilateral approach over a minimum period of three to five years. This can be seen as a direct and tangible outcome of the process of mediation in solidarity (*intermediación solidaria*) undertaken by FL, BL, and other European NGOs since 1993.

International and National Companies

Whereas the big-six banana companies—Chiquita, Dole, Fyffes, Del Monte, Pomona, and Geest—dominated the EU market until 1995, there are now just the first three. The level of capital-intensity and vertical integration required to make consistent profits in the banana trading business is such that concentration of ownership into fewer and fewer hands seems inevitable, especially given the weakness of most existing competition and antitrust policies. Charting the growth of Fyffes since 1993 is instructive: its rapid growth of EU market share from around 6 percent of a market of 3.5 million tonnes in 1993 to nearly 25 percent of a market of 4.1 million tonnes in 1996 has gone hand in hand with the acquisition—usually in fifty-fifty joint ventures—of a dozen major marketing, ripening, and distribution companies throughout the single market, from the Canaries to Scandinavia.

Apart from the big three, the only other successful actors in the EU market are national companies like Uniban, Banacol, Noboa, and Corbana (Table 5.1), which have contracts with their own national producers and have been busy trying to secure direct supply contracts with the major European retailing multinationals (such as Noboa with Sainsbury in the United Kingdom).

At the production end, though, trends are different. The transnational companies are tending to free themselves of direct ownership of plantations, in favor of guaranteed supply contracts with medium- and large-scale producers in the countries where they operate. This trend is not just confined to the banana sector, but it allows the Northern-based company headquarters to shift the responsibility for labor and environmental conditions on the plantations onto local shoulders, saying that these conditions are not in their control and that national legislation is in place to ensure that minimum standards are respected.

A 1995 report confirmed that since the Single European Market in Bananas came into operation in July 1993, the proportion of fruit traded by the multinational companies had increased from 51 percent to 58 percent in just thirty months (Arthur D. Little International 1995). But the impact of the alliance's multipronged efforts has altered the perspective and—in specific cases in particular locations—the practices and attitudes of at least the three remaining transnational corporate actors.

Governments and Intergovernmental Institutions

The interests and positions of the governments of most of the banana exporting countries and of the major European and North American consuming countries are, to say the least, diverse. On banana policy, conflict and lack of communication reign supreme. On matters of macroeconomic policy and practice, however, there is both little dissent and, in any case, little room for maneuver from the conventional neo-liberal wisdom. The accompanying box categorizes the main governmental actors according to their position on the EU's single market for bananas.

Generally speaking, the specific diplomatic and political positions of individual national governments can be derived by matching the groups of governmental actors with the five distinct types of trading arrangement that have evolved between the EU and exporting countries or between the EU and its domestic suppliers. To a certain extent, producer government interests and positions are also defined by the nature of the production systems that predominate in their own country. In the case of EU member states, their position still fundamentally relates to whether they have ex-colonial suppliers or domestic suppliers from overseas territories.

When a banana farmer with one acre in St. Vincent recently asked me in

Key Groups of Government Actors

- EU member states supporting 404: France, Spain, United Kingdom, Portugal, Greece, Italy, and Ireland

- EU member states against 404: Germany, Netherlands, Belgium, Denmark, and Sweden

- U.S. administration, especially the USTR and Latin American governments supporting the WTO complaint: Guatemala, Ecuador, Honduras, Mexico, and Panama (though not formally a WTO complainant)

- Latin American governments with guaranteed EU quotas: Costa Rica, Colombia, Nicaragua, and Venezuela

- ACP governments seeking bigger quotas: Cameroon, Côte d'Ivoire, Belize, Dominican Republic, and Ghana

- Caribbean governments not fulfilling quotas: St. Lucia, St. Vincent, Dominica, Jamaica, and Suriname

a strategy meeting why it was not possible to have a kind of producer cartel like OPEC as a way of moving beyond the current instability and ruthless market competition, it was necessary to explain the depth and complexity of the rifts that currently divide governments in the Caribbean Basin. The history of the Union of Banana Exporting Countries (UPEB) and its own multinational company, COMUNBANA, which the Latin governments set up in the mid-1970s, provides a cautionary tale. Essentially, the three big U.S.-based banana companies collaborated, with U.S. government help, to successfully "pick off" Latin governments and reduce and eventually get rid of the $1 export tax that had been collectively levied on each case exported from the region.

Since the 1970s, though, national governments' room for maneuver has been significantly reduced by other developments, particularly at the macroeconomic and macropolitical level. The key institutional actors at this level, especially the WTO, GATT, and the World Bank, have gained enormous influence in the macroeconomic sphere; as yet, we have little influence over these actors. Meanwhile, the United Nations Commission for Sustainable Development (CSD), set up to monitor the implementation of intergovernmental agreements signed during and since the United Nations (UN) summit in Rio in 1992, and the International Labor Organization (ILO), the UN agency mandated to defend multilaterally agreed on international labor conventions, both suffer from serious institutional weaknesses and a lack of

---

Key Intergovernmental Institutions

Commission for Sustainable Development (CSD), New York
World Trade Organization (WTO), Geneva
International Bank for Reconstruction and Development (World Bank),
    Washington, D.C.
International Labor Organization (ILO), Geneva
Food and Agriculture Organization (FAO)/Codex Alimentarius, Rome
International Network for the Improvement of Banana and Plantains
    (INIBAP), Montpellier

European Union Institutions

European Commission (DGs I, VI, VIII, and so forth)
European Parliament
Council of Agriculture Ministers
European Court of Justice
Economic and Social Council

---

credibility with the most powerful governments of the world. This has not, however, prevented the trade unions and other nongovernmental actors from seeking to engage with both institutions in an increasingly concerted manner.

The International Network for the Improvement of Bananas and Plantains (INIBAP), unique among the eighteen international agricultural research institutions in the Consultative Group on International Agricultural Research (CGIAR) system, is a research network, and it has a mandate to enhance smallholder banana and plantain production. This means that INIBAP's focus is on the nonexport sector, but it shares a common interest with the small farmers' organizations and plantation workers' organizations in seeking to develop sustainable production with improved varieties from a much more diverse genetic base. Collaboration between the European NGOs and INIBAP has a potentially fruitful future following an initial dialogue in spring 1996.

Both the UN Food and Agriculture Organization (FAO) and Codex play a critical role in relation to bananas: the FAO in hosting an annual conference of banana producing member states to analyze and evaluate world market developments, and Codex in developing and institutionalizing quality standards. It now also seems that the hardly transparent International Standards Organization (ISO) in Geneva may have a significant influence on the future of the banana sector, given its growing role in institutionalizing corporate standard-setting and environmental management systems.

Finally, there are the different institutional structures within the EU, which have all played—and will continue to play—a pivotal role in the evolving scenario. Each decision-making structure has required different strategies on the part of the alliance of nongovernment actors. These structures are in constant and complex interaction with all the companies and governmental and other intergovernmental actors and should be the subject of a separate analysis.

Principles and Practice in Mediating Sustainability

The Unifying Principles and Language of Sustainability

Since FL's first conscious mediation efforts in early 1993, we sought to match our intervention at the analytical level in the EU–ACP–Latin America banana trade with our overall objectives as an NGO: to promote a transition to sustainable agriculture and rural development (SARD) worldwide. The focus and locus of our work was now the East Anglian farming community, but the international perspective must continue to inform our own process of transition in the subregion, both in the rest of Europe and in the rest of the world.

Using a deliberately broad definition of the key elements of sustainability—economic, ecological, and social—FL sought to target its very limited resources on interventions that would bring key nongovernmental actors in the South and North *together* under the broad umbrella of sustainable production and trade. Then, in 1992, shortly after FL was established as an independent organization, bananas presented themselves as a pilot commodity to explore what this might mean in practice. Beginning with the first direct request from WINFA to mediate in mid-1992, we are now in a situation in which several hundred organizations are engaged in an increasingly irreversible process of transition toward a more sustainable banana trade worldwide.

Our experiences may well prove to hold pedagogical examples for achieving a transition toward SARD, especially in regions or subregions currently dominated by production for the world market. In my judgment, any alternative approach to a unifying one would have risked provoking more confusion, polarization, and conflict than already existed within and between organizations at that stage. Any resulting alliance or coalition of nongovernmental interests would have almost inevitably been partial or partisan.

In particular, recourse to the language and basic elements of sustainability has helped to emphasize during the whole complex process of information network and politicoeconomic alliance building that the social (especially the labor rights element) and the ecological are indivisible in our analysis and must remain so during our process of alliance building. The process of

Mediation as Intermediation plus Intramediation

In current English usage, "mediation" tends to imply the notion of intervention in a conflict in order to bring about a resolution. I would like to use the word more broadly. I see two distinct but overlapping types of mediation within the whole set of diverse mediations by individuals or organizations acting together in concert: intermediation and intramediation.

■ *Intermediation* encompasses a range of deliberate, or conscious, unilateral and multilateral interventions motivated by common educational, economic, and political objectives. *Educational mediation* can either be popular or academic; *economic mediation* can either be analytical or can involve becoming an economic actor in the sphere in which one intervenes; *political mediation* can be based around critique, or proposal, or both.

■ *Intramediation* is more about the separate but interlocking process of consciously gaining an understanding of the different actors and their interests within the evolving network and alliance. This aspect of the still ill-defined concept of mediation seems to be a prerequisite for mediation in solidarity.

With over three years' hindsight, a more practical definition of "mediation" might include the facilitation of exchanges between people and of information; the creation and nurturing of forums for informed debate, the construction of politicoeconomic alliances, the involvement of farmers and farm workers in research or in challenging existing institutions, or any combination of these activities.

alliance building, therefore, in itself, becomes a process of definition in practice of the loosely defined terms "sustainable banana production" and "sustainable banana trade."

The mediators in this process are all those active members of the alliance—in practice, individuals within mainly small organizational structures—who are keeping the ultimate objective of sustainability firmly in sight, recognize their respective viewpoints as Northern consumers or Southern producers, and are actively creating understanding in key areas of actual or potential conflict. In other words, they are intervening with the overall aim of construction of a well-defined but dynamically growing political and economic alliance around shared objectives.

Most organizations have more or less complex and sophisticated networks based on institutions in the shape of formal, semiformal, or informal

Chronology of Changes, 1992–96:
Four Years of Mediating Sustainability
in Solidarity

What were the decisive steps and activities that led, by mid-1996, to a situation in which alliances were strengthening, there was substantial interaction between the key actors and a range of economic and institutional actors at several levels, and alternative political, productive, and commercial proposals were emerging?*

March 1992: A WINFA delegation visited Honduras and Nicaragua and saw the plantation system for the first time. This visit came at a time when small-scale Caribbean farmers had heard rumors that their sole marketing company, Geest, had set up large new plantations in Central America. Governments knew and farmers suspected that if they were put in closer competition with Latin American plantations as a result of the changes to the import arrangements into Europe, they would be forced out of the market. But at this stage, Windward Island governments were telling their farmers that nothing was likely to change until after the end of their preferential trade agreement, which was legally enshrined in the Lomé Convention until the year 2000.

July 1992: During an evaluation of a two-week visit to the United Kingdom by a delegation of Caribbean small farmers, I was asked by WINFA to start sending regular information on the banana trade in the run-up to the decision over a single EU banana import regime. It also requested support in developing contacts with non-U.K. NGOs. Although WINFA had a funding relationship with two large British NGOs, it was felt that they were not fulfilling these important functions. FL's initial work on the banana trade therefore came at the specific request of the Caribbean farmers' organization.

September–December 1992: In conjunction with the two U.K. NGOs that had been supporting WINFA financially, FL gathered information from diverse sources in Europe, engaged in political advocacy according to WINFA's agenda, and facilitated a visit by a WINFA representative to

---

* Throughout the process charted here, focus was kept on external costs of production and the concept of sustainability. It therefore proved partially possible to draw previously divergent positions into a collective critique of the current production and trading conditions and to transcend the more partisan approaches that partial information and contact had led to in the past.

(cont.)

lobby U.K. and EU institutions, including at the Edinburgh summit. In December, the new regime was finally agreed on by agriculture ministers, setting a volume tariff quota on Latin American dollar bananas and tariff-free quotas for Caribbean and African imports.

May 1993: In the lead-up to the implementation of the new regime, FL took the opportunity offered by the inaugural conference of Via Campesina in Belgium to organize a meeting among WINFA, ASOCODE and ATC (Nicaragua) on the theme "Beyond the Banana Skin." Future exchanges between ASOCODE and WINFA were proposed, and the ASOCODE leadership committed themselves to developing a specific position on the EU banana trade. ATC reported that delegates from fifteen banana trade unions had just met at the regional level in Central America for the first time and were embarking on a process of common strategizing and political coordination.

October 1993: FL invited representatives of the Coordinadora de Sindicatos Bananeros from Costa Rica and of WINFA to join a delegation to the United Kingdom, Belgium, and Germany. This two-week visit proved crucial in dispelling many of the more simplistic analyses of some European NGOs, as well as cementing a direct relationship between banana workers' organizations in Latin America and small farmer organizations in the eastern Caribbean.

February 1994: FL organized the first meeting of European NGOs to focus on fair banana trade. The eastern Caribbean ambassador to the EU updated the participant NGOs and NGO networks on the banana wars in GATT and in the actual market. A pan-European approach, in full collaboration with workers' unions and farmers' organizations, was endorsed.

May 1994: Alter Trade Philippines, a successful alternative trading company working with producers in Negros and Alter Trade Japan, an offshoot of the large-scale Seikatsu Club Consumer Cooperative, visited the United Kingdom and Germany to report on their successful alternative banana trade directly with Japanese consumers.

July 1994: A common statement denouncing Geest's role in the violent strike of May 1994 in Costa Rica and calling on the EU to establish mechanisms to encourage sustainable production and trade was sent to all members of the European Parliament. The statement was signed by trade unions in Latin America and by the

(cont.)

International Union of Food-workers; by farmers' organizations in the Caribbean, Central America, and Europe; and by twenty European NGOs.

September 1994: WINFA invited FL to the Windward Islands to help facilitate the visit by an ASOCODE delegation from Belize and Costa Rica. The work of European NGOs and the links with Latin American unions were presented to Caribbean farmers and NGOs. FL also visited the WINFA member organization OPAM, in Martinique, to assess and report on the situation facing small producers. The language barrier had prevented English-speaking representatives from finding common cause among banana producers in the neighboring French-speaking islands.

November 1994: Solidaridad held a tribunal on the EU banana regime with witnesses from the Caribbean, Latin America, and the Canaries. Small-producer representatives from Ecuador and Dominican Republic met with the group of European NGOs working on fair trade.

December 1994: The French NGO RONGEAD organized a meeting with an Ivory Coast farmers' representative and European NGOs to learn about the situation for small producers and to report on the emerging alliances in Europe, Latin America, and the Caribbean. RONGEAD representatives also visited Cameroon to identify organizations working on banana issues there. A proposal to organize a strategy meeting for small exporters and unions with potential productive projects emerged.

March–April 1995: A representative of the women workers' section of ATC was invited to the Social Summit, to an international conference on sustainable agriculture and rural development in Norway, and to update German NGOs on progress with the workers' cooperative project in Chinandega. A WINFA representative visited FL to help evaluate the progress of our Education for SARD program in East Anglia. In this way, the relationship with WINFA has become genuinely two way.

May 1995: The third conference of Latin American workers' unions met in Honduras, with the Ecuadorean small producers' organization UROCAL and European NGOs participating. Full consultation about the proposal for a preferential fair-trade quota was sought from the unions.

(cont.)

June 1995: European NGOs, having drafted a rigorous critique of the continuing lack of sustainability in European policy, agreed to launch a common campaign in support of the fair-trade proposal in autumn 1995. Also, WINFA and ASOTRAMA toured Germany and Switzerland.

July 1995: A strategy meeting between the EUROBAN coordinator and Coordinadora in Nicaragua was held to set in motion a process of defining and implementing labor and environmental clauses—defining minimum standards—in future banana trade agreements.

October 1995: Coordinadora toured nine EU member states and Brussels institutions in three weeks, starting at an international meeting.

February 1996: WINFA launched interisland mobilization efforts around farmer ownership of WIBDECO, which now controlled 50 percent of the profits of the former Geest banana trading division, operating with a guaranteed 78 percent of the total exports from the four islands. The Windward Islands still, collectively, underfulfilled their allocated quotas by over 30 percent in 1995. The islands' farmers faced a historic crossroads, after an expensive partial buyout of their own banana industry by the governments, in collaboration with Ireland-based Fyffes.

April 1996: EUROBAN-WINFA-Coordinadora developed a common political position on 404, fair trade, and socioecological clauses for all trade, and agreed in principle to drafting a common work program and budget for 1997 to 1999.

groupings; individuals; or any combination of these. It is only natural that the coordinator should recognize, collaborate with, and actively seek to enhance this natural decentralization of the majority of network- and alliance-building activity going on in the name of sustainable, fair, or just "green" bananas.

It is also to be expected that many of the details of interactions in the networking process are unknown to the coordinator. In nurturing a broad-based international network, it appears to be necessary to spread the control over the practical evolution of common or coordinated activities. In this sense, power must be shared as widely as possible among the actors closest to the core of the process. The coordinator must not be tempted to play the role of information power broker, as this would rapidly undermine the integrity of the whole process.

Alliance Building with Information as the Lifeblood

The objective was to give credibility both to the mediation initiative—in solidarity and through the optic of sustainability—and to the process of networkbuilding itself. Information flow was put center stage in the toolkit of strategies deployed by the coordinator and other key mediating forces within and among EUROBAN, FL, and WINFA. Without an independent and trustworthy capacity for information gathering, processing, analysis, publication, translation, and dissemination, the efforts to build an alliance based on a perceived common economic and political interest in stable, fair, and healthy market conditions would have foundered from the outset.

It has been essential to build up the capacity to publish every two or three months a bulletin of banana trade news. Ultimately, this processed information is becoming more and more valuable market intelligence for different economic and institutional actors. This in itself can become a significant factor in shifting the balance of power within the international banana economy. Managing and channeling the flow of information, which grows as the networks and alliances grow, is a central role for the coordinator or facilitator. Especially at a time of burgeoning international contact using new technologies, the information function of EUROBAN in particular, as well as of both WINFA and the Coordinadora, has become difficult to manage, first because of its sheer volume, and second because of the growing complexity of the jigsaw puzzle being pieced together by forces seeking to change the status quo.

The growing volume—and the sheer speed—of information flow now, compared with that of even five years ago, is remarkable. The network is not necessarily able to cope with the increased information flow immediately, because the capacity of its members continues to lag behind the technological potential. As the international networking and the alliance's common program evolve over the next few years, it will be important to study the dynamic role of information itself as well as different information media.

The critical issue remains that a network for banana sustainability and solidarity that challenges the commercial and diplomatic status quo stands or falls by the quality and timeliness of the information available for internal and external digestion. But to be most effective, such an information function would require a greater level of sustained funding for a minimum of another seven or eight years.

Beyond Consultation to Equitable Participation

The tasks of short-term mediation in the first few years of building a genuinely international alliance have been moving toward securing a longer-term process of change. This has proved—and will continue to be—critical to mediate successfully in the process of democratization of the whole set of relationships between European NGOs, trade unions, and farmers' organizations on the one hand, and the Latin American and Caribbean trade unions, farmers' organizations, and NGOs on the other. The mode for most European organizational relations with counterparts or partners in the South has often been consultation without adequate respect for and recognition of internal democratic processes in the organizations with which they have developed a bilateral relationship. The rhetoric is frequently of equitable participation, but the practice is all too often twofold: insufficient information flow, and a lack of real participation in decisions by genuinely representative organizations in the South.

The whole practice of democracy within the network- and alliance-building processes therefore becomes critical in levering a shift in attitudes and practices toward a more equitable approach within our own European-based organizations. The transparency of meetings, structures, finances, and the collective decision-making process that leads to concerted action must never be compromised by unequal participation. The mediator's overarching responsibility in this situation is to ensure that the different (sometimes divergent) voices are heard and are taken into account within the networking process.

Consumer Education, Action and NGO Diplomacy

We began with a few relatively weak and atomized efforts around Europe to promote education and critical debate about banana production and consumption before the single market. Three years later, we have successfully undertaken a joint consumer postcard-signing action in six countries and have coordinated consumer education about the banana trade and collective action on wider development and agricultural policy issues.

In this respect, the key task for FL was to mobilize in Britain the major development NGOs, other NGO networks, individual consumers, and, where possible, trade unions. This mobilization essentially took place around participation in the U.K. lobbying work for market access for fairly traded ecological bananas. The concrete result was over 10,000 signed postcards and 300 letters to Britain's minister of agriculture and to the EU's Commissioner Fischler. Together with cards from parallel consumer actions in Holland, Germany, Denmark, and Austria, the weight of pan-European opinion (over 150,000 signed postcards) has certainly helped generate a direct and constructive dialogue with the highest-level EU decision makers. But the civil servants' hands remain "tied by GATT/WTO,"[6] and by the de facto political paralysis of the Council of Agriculture Ministers.

This European postcard action, in conjunction with the "Yellow Fever" proposals endorsed by EUROBAN, was the first tangible collective action of the European NGO alliance, launched in October 1995 and delivered in March 1996. The action was quite different in each country. In the United Kingdom, it was the product of NGO network mobilization, predominantly by church-linked and student groups, as well as public signings in the street. FL also acted as a catalyst to a successful consumer education and action program in Ireland, where the Irish Fair Trade Network is now coordinating diverse activities around the banana trade.

The coordinator has also been proactive in catalyzing consumer education and action in Norway following visits in early 1995 and May 1996. A joint consumer action at a conference in Roros in March 1995 generated a comprehensive environmental mission statement from Dole's lawyers over a year later. This document gives the alliance the basis for independent monitoring of Dole and opens the door for possible engagement by EUROBAN actors in the future.

Now that a "fair trade" labeled banana is known to be in demand in U.K. and Irish markets, our capacity to negotiate with economic, governmental, and intergovernmental actors has been greatly enhanced. Together with selected economic actors from the producing countries, we are now in a position to start exerting concerted pressure on Europe's only remaining major private-sector banana trade actor, Fyffes, to cooperate on equitable terms at both economic and political levels.

Yet the stage of maturity now being reached in this example cannot hide the fact that there are residual conflicts of vision or analysis between different NGO actors in Europe. For example, the persisting feeling in the Spanish, French, and Portuguese overseas territories that the northern European NGOs would be happy to see their "domestic" production disappear explains the continuing fragility—or even nonexistence—of banana consumer education or action initiatives in these countries.

New Institution Building and Capacity Strengthening

Three parallel processes of new institution building and capacity strengthening have been unfolding autonomously, and converging, since 1993, as summarized below. Taken together, they represent the emerging basis of long-term cooperation over sustainable production and trade in bananas.

Coordinadora de Sindicatos Bananeros de América Latina *Building:* A new regional coordination structure and two new national coordination structures have emerged. Besides unifying all the independent unions in the seven major exporting countries of Central and South America, contacts are now being developed by the regional structure in the Philippines, Belize, Venezuela, Suriname, and the Windward Islands. The detailed proposal from the regional Coordinadora for social and environmental "clausing" (definition of clauses) is now widely known by key European governments and European intergovernmental actors.

Strengthening: The organized banana workers' capacity to analyze, denounce, propose, and cooperate with other actors has been massively increased, largely through efforts unsupported from outside the region. They are currently developing strategies for the year 2000 and beyond, centered around the implementation of minimum labor and environmental standards in all export banana production and the establishment of worker-based and small-farmer-based enterprises. This new energy and direction have empowered the trade unions in their struggle against undemocratic forms of worker organization, such as *solidarismo*.

WINFA *Building:* New independent banana farmer structures in all four islands have emerged since 1993, with a strong mediation role played by WINFA's program officer. Apart from the quest for economic survival, since the beginning of 1996, the independent banana committees face the challenge of how to organize farmers to gain democratic economic control over up to 50 percent of their own industry. Such a concept of farmer ownership will not come about without a high level of organizational discipline and a will to rebuild local banana farmer structures to replace the largely discredited island-based growers' associations.

Strengthening: Negotiations with the Caribbean banana exporting companies and European NGOs to supply a "fair trade" labeled banana in the United Kingdom and Ireland have now thrust WINFA into a strong role of leadership in the quest for a realistic farm diversification program in general, and in the process of implementing environmental improvement programs with banana farmers in particular. This should lead to a strengthening of WINFA's organizational capacity for its proposed five-year diversification program and could avert a conflict with the proposals for restructuring the banana sector made by British and EU-funded consultants in 1995.

EUROBAN *Building:* The synergy generated by increasingly overlapping collaborative efforts within EUROBAN, and between EUROBAN and other actors, means that we have built a resilient but as yet entirely informal structure for NGO cooperation and coordination. The test will come as we democratize our multiple relationships with our key Southern partners. A key linkage that will prove crucial to sustained efforts in Europe is the recent progress in involving European-based trade union confederations in the work of EUROBAN.

Strengthening: Both the reduction of duplicated efforts and the process of consensus building required to produce common analyses and alternative policy proposals have strengthened most NGO actors appreciably. Most have secured independently generated finances (mainly at the national level) to work on elements of a common program at the pan-European level and beyond. Our capacity to negotiate EU funding is greatly enhanced by the level of consensus now being achieved over future strategies. We have also strengthened our links to other pan-European NGO networks working on agricultural and trade policy issues.

Toward a Common Trilateral Program?

At the last EUROBAN strategy meeting in Denmark, the coordinator and the French NGO partner RONGEAD were mandated to prepare, in full cooperation with WINFA and the Coordinadora, a three- to five-year common work program and global financing plan. This is not to replace efforts to develop and to become self-financing within our own regions, but rather to complement them at the international level—the level at which our governments are failing to have any dialogue. It remains to be observed, evaluated, and analyzed what will happen to this stated intention to develop a new trilateralism that can act simultaneously at the political, economic, and educational levels. It will also be interesting to chart the response of the bigger European nongovernmental development agencies in the first stages of the presentation of such a dynamic, tripartite approach to the medium-term future. Perhaps the only certainty is that the outcomes of this approach are not easily predictable.

A Transition Toward Long-Term Sustainability?

In the economic sphere, we have all, collectively, lacked sufficient—and sufficiently solid—examples of what we are talking about when we refer to sustainable production for the world market. We have therefore been structurally weak in the process of proposing and constructing a different kind of banana trade. Until very recently, this fragility has seriously limited our capacity to engage in sustained interactions with most of the significant economic, governmental, and intergovernmental actors involved in the world banana trade. Over the last year, as the official European political process surrounding the management of the banana trade has become well and truly paralyzed, the fragility of our common position has decreased as our credibility has increased. We have become serious actors in a fast-changing scenario.

But if a political vacuum has opened up at certain institutional levels, making them relatively easy to move into, in the economic sphere, power relations remain relatively unchanged. Accelerated concentration of the market has meant fewer and more powerful economic actors with whom to seek dialogue, engagement, or confrontation. However, such was the level of concertation by mid-1996 that it was becoming possible to envisage much more rapid progress in the economic sphere as a more tangible common front could be demonstrated among critical consumers, workers, and family farmers. More practical evidence is still required to catalyze more than marginal economic changes.

There are, of course, banana worker and farmer cooperatives, but they are few, weak, and under almost intolerable pressures from the key economic and institutional actors in the conventional banana trade. There is ecological production in many countries of tropical Asia, Africa, Latin America, and the Caribbean, but in very few is this production destined for the export market. There are at least four or five examples of alternative trading arrangements for bananas and banana products worldwide, but they are fragile and continue to risk commercial, political, or even physical sabotage. "The threat of a good example" seems to apply especially eloquently when it come to the history of attempts over the last two decades to establish models based on different visions of development in the banana sector.

However, the key institutional actors, who have essentially defined the economic and political parameters of the game thus far, are even more diverse in their objectives and strategies than are those in the nongovernmental sectors. In reality, they have substantively divergent interests in the banana market. Therefore, the marketing companies or the governments have not pursued any collective strategies in the face of either the new Single European Market itself or the growing coherence and concertation among those trying to change the economic and political power relations within the international banana economy. Like any NGO, we have been able to engage them in dialogue or correspondence at very different levels. This engagement, however, has occurred essentially bilaterally or by maintaining multilateral relations among NGOs, trade unions, and farmers' organizations, often with other actors from civil society coordinating their approach to the institutional actors in and from several countries.

European Reflections on Mediation Strategies

One clear reflection from a network coordination perspective is that unless exchanges of substance between third parties within or at the immediate periphery (leading edges) of the network are fed back to the coordinator—or indeed to the wider alliance through EUROBAN, WINFA, or Coordinadora strategy meetings—neither the coordinator nor the semiformal EUROBAN forum may be in a position to mediate effectively and constructively at the many levels required; the result could well be the failure to develop and implement a common advocacy strategy and common work program across several regions of the world.

Another clear and critical observation is that the coordinator (chief mediator) should ultimately defer to the greater experience—in many activities and on many issues—of other individuals and organizations, and not just to those who are directly involved in export banana production in Latin America or the Caribbean. For example, a group in Switzerland has been trying to change the terms of the banana trade since the early 1970s; others in Holland, Belgium, and Germany each have seven or eight years' experience in solidarity trade or fair-trade arrangements with small national producers or state-owned plantations in Latin America. Their cumulative networks of contacts, as well as their evaluation of years of detailed intervention and mediation processes of their own, mean that the coordinator should pay particular attention to their active participation in all decision-making processes within the European regional network.

At another level, EUROBAN collectively has to defer to the greater experience and legitimacy of the banana organizations in the Caribbean and Latin America: the trade unions, the farmers' organizations, and some NGOs. They are more vulnerable than the European organizations to the consequences of diplomatic errors or miscommunication, for example. A very real danger is that the speed of European and international political and economic developments in the sector has been so great that European NGOs can sacrifice real consultation and negotiation on the altar of meeting imposed (and self-imposed) deadlines. The rhetoric of solidarity can quickly be proved to be empty on occasion.

These are perhaps harsh reflections at a juncture where the alliance's economic and political influence looks set to grow significantly. However, critical self-evaluation becomes increasingly necessary to counter the tendency to be swept along by real events. For FL and BL, the process of development education and the process of real social, ecological, and economic change have become inseparable, and the need for constant analysis, evaluation, and task setting demands ever-higher levels of equitable consultation in key areas of what might be called common network development.

Latin America: *Nuevo Sindicalismo* and Social Enterprise

Many banana workers' trade unions were close to collapse by the early 1990s, especially in Costa Rica, where the attempts to impose *solidarismo* as an alternative model for employer-employee relations originated. There are many bitter experiences of repression and injustices against banana workers, especially those who persisted in defending the right to free association and independent organization. But, led by the Costa Rican unions, banana workers across the region are now creating a new model of trade unionism for the twenty-first century. They see themselves as engaged in a struggle to renew unions as future-oriented social institutions. The success of the unification of banana unions in seven countries in the region since 1993 has been crucial to the confidence with which the trade unions are pursuing distinctively new strategies.

The emphasis within the union movement is now increasingly on (self-) education for participation in all those decision-making structures that affect workers directly and indirectly. It is crucial that the grassroots membership shares the leadership's vision of the new unionism; otherwise, the unions' role as important institutions in civil society will sooner or later be undermined.

Detailed proposals from the regional Coordinadora for social and ecological clauses to be written into all future banana trade with the EU are forcing the European NGOs to reevaluate their longer-term political strategies. We are forced, for example, to clarify the distinction between "conditioning" all trade with minimum labor, health, and environmental standards and making efforts to find producers who are prepared to meet higher social and environmental criteria in exchange for a premium price. Clausing and fair trade can be complementary advocacy strategies, but it was the unions that propelled European NGOs into an important debate about political and alternative marketing strategies.

Although there are good examples of social enterprises in the forestry sector in Central America, banana workers still face a real struggle to turn their plans for worker-controlled and worker-owned enterprises into reality. The success—in the face of virtual economic collapse—of the 4,300-member Nicaraguan workers' cooperative Trabanic in creating alternative employment is unfortunately an isolated example. The cooperative at El Trianon in Chinandega province still cannot plant its ecological banana plantation because of continuing insecurity of tenure. But the unions recognize that, in the medium to long term, they will have to become direct promoters of democratic economic initiatives, if not economic actors in their own right.

On the question of sustainability, the plantation workers' unions tend not to include the socioeconomic elements that we Europeans often include implicitly or explicitly. They tend to describe the elements of their alternative vision as ecological sustainability plus social justice or equity. This issue of language and usage is going to become increasingly important as the nongovernmental actors seek to work in closer cooperation.

Windward Islands: Diversification and Farmer Ownership

In the Windward Islands, the key debate over the future economic survival of 25,000 small-scale banana farmers revolves around a different word— "diversification," both within the banana sector (other banana-based products and processing) and outside of it. For WINFA, the strategy is to reduce farmers' dependence on the monocrop while seeking to enhance local and regional food security by more domestic food production. WINFA's own five-year farm diversification program, designed in full consultation with farmer members on five islands, includes environmental improvements in the banana sector as well as a practical transition toward more sustainable farming systems for those who choose to move out of the increasingly vulnerable export crop.

The land reform, which saw many former plantation workers in the islands become small-scale farmers, has largely been completed. Compared with many other examples around the world, land reform in the Windwards has been relatively successful, even if the export monocrop has taken much greater hold since land reform. It is the economic basis on which achievements in human and social development have been made, especially in the 1980s, even though the producers' share of the profits remained low and outside the control of farmers themselves.

But as of the end of 1995, the Windward banana farmers have been thrust into a new phase of their industry's development. WIBDECO entered a fifty-fifty joint venture with Fyffes to buy what was Geest's banana business. For the first time, this gave farmers a chance to gain a stake in the shipping and marketing of their own fruit. But a critical political issue surrounds the future of the growers' associations themselves: most farmers now see them as having lost credibility and no longer being representative of their interests. In this situation, independent banana farmer movements, often nurtured and supported by WINFA, have emerged in each of the islands. It remains an open question whether the potential to secure farmer ownership of WIBDECO will mobilize a new, more democratic banana farmer movement, and vice versa.

In this process of change, the mediation efforts of WINFA leadership will surely prove decisive in focusing the energies of independently organized farmers on sustainability in the long term.

From Fair Trade to Sustainable Trade?

It is really the European NGOs that need to ask themselves about the issue of development models. Our "partners" are increasingly clear: social justice, ecological health, and economic democracy are the inseparable basis of sustainable development. In the European arena, these elements are all too often disaggregated; separate movements exist around each of the elements. It should be no surprise, therefore, that a fair-trade movement focused on social development for marginalized producers in a given tropical commodity sector does not necessarily cooperate with distinct movements focused on ecological protection and improvement or on economic democratization.

The NGO response to the sustainability agenda is still relatively atomized and weak. The quest for a new type of solidarity between consumers and producers is very much in its infancy, and it risks remaining so, as long as the debate over the future of international trade remains largely intellectual.

It is surely possible, however, to use the language of sustainability to bring together a common struggle for sustainable trade, integrating the three key elements of social justice, ecological health, and economic democracy into efforts to match sustainable production and sustainable consumption. The example of the fruits of mediating sustainability in the banana sector underlines the importance of such an approach, at least in the initial processes of interregional network development.

It remains to be seen whether this linguistic device remains useful as we move into a new phase of developments in the banana sector. Having created significant political space in a specific commodity sector, the question of what economic space can be created in practice becomes central. Without such "live" examples, talk of a new international economic order is likely to remain in the realm of abstract theory or speculation. Only conscious and concerted attempts to change the terms of trade will lead us to a better definintion of sustainable trade.

The trade and fair-trade world has changed rapidly since the completion of this chapter in mid-1996. An additional note is required, therefore, to put the story told so far into a new context.

The new thrust by the United States, the EU, and other OECD countries to sign a radical new agreement on private international investment rights (but not responsibilities)—the Multilateral Agreement on Investment (MAI)—threatens to further undermine hopes of a transition toward socially, economically, and ecologically sustainable development worldwide. The World Trade Organization is flexing its legalistic muscles in the global economy, which is quickly dividing into regional blocks. In the case of the banana dispute, one of the big corporate players, Chiquita Brands International, has fought hard—and controversially so—to win a settlement generally seen to be in its favor.

Meanwhile, the regulatory framework for national governments as well as for international intergovernmental institutions (such as the ILO, FAO, and WHO) and regional intergovernmental institutions (such as the EU, NAFTA, and ACPEC) is increasingly being defined by trade and economic policies that promote a less regulated and less accountable environment for private (foreign) investors. Private investors (mainly multinational companies) now spend a record quarter of a trillion dollars per year on expansion outside their country of origin.

The world banana economy saw some of the lowest prices ever in the mid-1990s. Only climatic events such as hurricanes, flooding, and El Niño have kept prices from falling even lower. In the EU, the only major market with border restrictions, prices have also been historically low, although still well above world prices. The partial liberalization of the EU's controversial

1993 single market regime has already brought many small- and medium-scale producers to the brink of survival, some 15,000 to 20,000 having been squeezed out (EUROBAN/Solidaridad 1995). Thousands of plantation workers from Honduras to Ecuador and from Colombia to Belize have suffered a fall in living standards, working conditions, and job security, and tens of thousands have been laid off permanently, despite the expansion of export banana hectarage.

In the face of this increasingly harsh socioeconomic reality, some limited improvements in environmental practices have taken place in some countries, often in response to perceived pressure from European and North American consumers. Certified organic banana trade now registers in market-share figures. Additionally, fair and alternative trading initiatives have grown since 1996 to involve producers and consumers in Ghana, Ecuador, Dominican Republic, Costa Rica, Windward Islands, and the Philippines and in six European countries (in Switzerland, such initiatives supply 14 percent of the market) (FLO Banana Register 1997, personal communication). By mid-1998, organic, "fair trade" labeled, or alternatively traded fruit is projected to account for 2 to 3 percent of the total banana market of the European economic area.

This latter market impact—together with increasing consumer pressure and media attention in key countries such as Germany, Britain, and Sweden—has more or less obliged four of the five major corporate players to engage with NGOs, trade unions, and farmers' organizations on at least part of our core agenda. This is taking place at different levels with different banana companies, but concerted efforts center on bringing these key players as well as key governmental players to the negotiating table of a major international banana conference in May 1998.

The agenda of such a process of preparing and staging an international event is clearly defined in the title: Toward a Sustainable Banana Economy. Many of the policy mechanisms for achieving a transition in this symbolic sector of the global agricultural economy are now being tried and tested; some remain to be developed, such as sustainability auditing, harmonized fair-trade standards, tariff recycling, sustainability clauses in trade agreements, even the first multilateral (sustainable) commodity agreement. At the very least, the international banana charter that will come out of this process should be seen as evidence that the mediation of sustainability and solidarity backward and forward along a specific economic chain can bear fruit when it comes to unlocking the potential for a qualitative shift in the global banana economy.

Notes

1. Nearly 90 percent of internationally traded bananas are of the Cavendish variety.
2. My own calculations, based on company and market price data.

3. Inter-Press Service, Panama City, 10 February 1995.
4. Known as regulation CoM 404/93 (Brussels).
5. Press conference, San José, Costa Rica, 30 March 1993.
6. Personal communication from Commissioner F. Fischler to Solidaridad, 30 November 1995.

References

Alternative Treaty on Trade and Sustainable Development. 1992. International NGO forum, Rio de Janeiro.

Arthur D. Little International. 1995. *Study of the Impact of the Banana CMO on the Banana Industry in the EU*. Paris: ADL Consultancy.

ASEPROLA. 1993. *What Is Solidarismo?* Costa Rica: ASEPROLA.

Banana Trade News Bulletin. 1994–96. nos. 1–7. Farmers' Link/Banana Link, Norwich, England.

Bergquist, C. 1986. *Labor in Latin America*. Berkeley, Calif.: Stanford University Press. Cited in T. Fenton and M. Heffron, eds. 1989. *Transnational Corporations and Labor: A Directory of Resources*. New York: Orbis Books, 23.

Biekart, K., and M. Jelsma. 1996. *Peasants Beyond Protest in Central America: Challenges for ASOCODE, Strategies for Europe*. Amsterdam: Trans-national Institute.

CASA (DK). 1996. Real Bananas. Report of DANIDA seminar, Copenhagen, April.

Coordinadora. 1996. Sindicalismo hacia el Siglo XXI. Coordinadora de Sindicatos Bananeros de Costa Rica, Siquirres, Costa Rica, February.

EUROBAN/Solidaridad. 1995. *Yellow Fever: Proposal for a Specific Fair Trade Banana Allocation*. Utrecht: EUROBAN/Solidaridad.

Farmers' Link. 1995. *¡Just Green Bananas! UK Campaigners' Guide to the Banana Trade*. Norwich, England: Farmers' Link.

Hines, C., and T. Lang. 1993. *The New Protectionism: Protecting the Future Against Free Trade*. London: Earthscan.

ICTUR. 1995. Proposal for the Social and Environmental Clausing of Banana Trade. Coordinadora de Sindicatos Bananeros de America Latina, Chinandega (Nicaragua), September 1995. Summarized in International Union Rights *3 (1)*.

International Water Tribunal. 1992. Judgment and Recommendations in Case Against Standard Fruit Company (Dole) in Costa Rica. Amsterdam, February.

Oxfam (UK). 1996a. Submission to Parliamentary Environment Select Committee Trade and Environment Hearings. Oxford, January.

Oxfam (UK). 1996b. Reforming World Trade: The Social and Environmental Priorities. Oxford, June.

RONGEAD. 1995. *Bananas: Development Dynamic or Dynamite?* Trade Briefing no. 1. Lyon: RONGEAD.

Solagral. 1994. *Commerce: Quoi Qu'il en Coute?* Paris: Solagral/FPH.

Solidaridad. 1995. Position Paper Concerning the Report on the Operation of the Banana Regime of the EU. Utrecht, November.

UNED-UK. 1994. Reducing Britain's International Agricultural Footprint. In *Round Table Report on Land, Sustainable Agriculture and Rural Development*. UNED-UK NGO submission to U.K. government and CSD, September.

6

Dealing with and in the Global Economy

Fairer Trade in Latin America

Pauline E. Tiffen and Simon Zadek

The last decade has seen the consolidation of a theory and practice of economics that has been loosely, and with ironic inaccuracy, called "free trade." People throughout both the developed and the developing world have forcibly adjusted and made profound efforts to live up to the new conventional wisdom. But despite the promises of material reward for all, this new economics has meant that most people have to work harder for less immediate reward. In Mexico, for example, the majority are poorer than they were a generation ago; in Nigeria, incomes have fallen by a quarter since the mid-1970s; and in other places, the facts are worse ("Mystery of Growth" 1996). People have waited patiently but often in vain to gain the benefits said to accrue from a global economy supposedly characterized by new opportunities in open markets. An economics that recognizes need only if it is backed by purchasing power does not effectively avoid the negative social and environmental consequences of the economic adjustment that have been well documented by many bodies, from environmental nongovernmental organizations (NGOs) to the World Bank.

Although these negative views of the effects of the process of globalization are correct, far less has been written about the "openings" afforded by such fundamental change and the new concepts arising from the adjustment process—the varied struggles not just to cope with liberalization and its consequences but to make the most of it and to overcome or outdo it. This chapter aims to outline the efforts of a gamut of organizations that have attempted to deal with and alter the accepted norms of the new global economy. Even as barriers have tumbled down and intervention in the market has been all but outlawed as protectionist under the new World Trade Organization (WTO) regime, a fair-trade movement started in the 1960s has begun to gain momentum. The exponents and champions of this fair trade explicitly espouse and flaunt noneconomic

values and purposes for trading and seek as a point of fundamental principle to see people as the ends and not the means of economic activity.

The fair-trade movement and the alternative trading organizations
(ATOs) that are behind it are offering a political concept and practice of a
fairer approach to business. The experience and reach of the movement are
wide ranging. It is active in all continents, full of spectacular achievements in
penetrating markets and overcoming booms and busts, and gifted with the
self-sacrifice and heroism of inspired individuals operating in countries with
oppressive regimes. But, academically speaking, the experience is scantily
documented and thus does not yet, arguably, qualify as theory. Like many
social movements, it is a de facto reality yet to be acknowledged by the de
jure processes and formal contexts of intellectual thought.

One of the few books written explicitly with a view to describing the origins and progress of the movement has been Michael Barratt Brown's *Fair
Trade* (1993).[1] Much has changed since Barratt Brown published his book.
The movement has grown in size and in many countries has gained a higher
public profile. Within technical and development cooperation programs,
NGOs, governments, and multilateral agencies alike are embracing the strategy of supporting small business growth, in keeping with the ideological
framework of a global market. In this context, fair trade as a development
strategy has for many traditional agencies come to represent a reasonable
blend of market-based economy and social justice and environmental interests. This is particularly the case for those disillusioned with the failure of
most aid-driven development processes.

Alternative or fair trade is small within the context of overall trade
between the North and the South, between developed and developing
countries. It is probably between US$300 million and US$500 million in
retail sales value per annum, or about one hundredth of 1 percent of the
annual world trade of US$3.6 trillion (Cavanagh 1995, quoted in Niessner
1996). But this is not the whole story of the significance of alternative
trade. Although not shying away from values and volumes as indicators,
this chapter aims to push beyond using the conventional measurement of
wealth (as a contributor to development), in keeping with the wider trend
of incorporating previously excluded externalities into conventional statistics (MacGillivray and Zadek 1995).[2] To do this, we examine in some
detail the philosophies behind alternative trade and the threats of free
trade to livelihoods. We focus, however, on some different organizational
experiences of small-scale farmers of primary commodities—coffee and
cocoa—in dealing with the free market, and on alternative approaches to
creating environmentally sustainable production for artisans living in rainforest zones.

What Is Known About Alternative or Fair Trading?

ATOs have traded a diverse range of products over almost three decades, with, for many years, a strong emphasis on handicrafts and textiles (Table 6.1), and with production based on traditional methods and techniques.[3] Traditionally, alternative trade in crafts and textiles has been small scale, with strongly reasoned producer partnerships that have tried to offer opportunities to the most marginal of the rural and urban poor—women, the disabled, those involved in cyclical work (harvests)—against the particularly harsh and unremitting backdrop associated with these sociological groups and occupations in the developing world. Trade in crafts and textiles has also been a springboard for consumer education: about other cultures, wider economic development, world trading regimes, and tariff discrimination issues (for example, the Multi Fibre Agreement).

Much of the earlier experience in fair trade was in so-called ethnic crafts and household decorative items. Indeed, the movement played a significant role in introducing Northern consumers to these products. In the 1990s, however, only a fraction of traders in these markets have taken on any social emphasis or responsibility to care for the conditions of suppliers in the products' countries of origin. As such, this traditional ATO activity— with total sales of around US$70 million in 1993–94—finds itself under constant and increasing competition.

The most rapid growth of fair trade in recent years has been in food products, particularly coffee. This growth, built on twenty years' work in building consumer awareness and solidarity, was initially fueled and accelerated by the Max Havelaar Foundation (MHF) in the Netherlands. In 1988, the MHF established a set of criteria to which conventional companies could adhere and receive an independent mark or seal of approval from the foundation.[4] Among the criteria for the fair trade of green coffee, for example, was a floor price paid to producers below which a company using the mark should not fall, regardless of the prevailing market price. Since 1988, almost US$20 million has been paid to MHF-selected producer supply organizations just from the fair-trade "surcharge" on products—that is, the differential between the market price and the price paid for coffee sold with the Havelaar seal of approval (Ransom 1995b).

Another critical factor determining the returns to small producers of coffee has been the source of the commodity and the terms of trade. The criteria developed by the MHF went to the heart of the ancient dilemma of the weak and atomized position of most small-scale farmers, artisans, and other primary producers in relation to local middlemen or market makers. Under

Table 6.1 Craft and Textile Sales via Top-Ten Alternative Trading Organizations, 1993–94

Country	Organization	Value (US$ million)
UK/Ireland	Bridge/Oxfam	13.5
Australia	CAA Trading	10.4
USA	Selfhelp Crafts	8.5
Germany	Gepa	5.6
Germany	Eine Welt/Team	5.4
USA	SERRV Handicrafts	4.4
Netherlands	Fair Trade Organisatie	4.3*
Austria	EZA	3.4*
UK	Traidcraft	3.1
Italy	CTM	2.2

Source: Adapted from Bridge. 1996. *Bridge Framework Manual,* p. 81. Oxford: Oxfam.
*Figures are from 1992–93.

the MHF scheme, all qualifying coffee is purchased directly from small-scale producer organizations and exported by farmer-owned or -associated export companies. Affordable prefinance is part of the deal and must be made available by the buyers to end farmers' dependence on local moneylenders and to keep their coffee out of the hands of often unscrupulous exporters.

The MHF message that viable trade can be underpinned by a fairer deal to the economically weaker partner—the small producer—has found increasing resonance in all industrialized markets. Coffee was a fitting traditional commodity for the movement to work with, in that small-scale peasant family farms are a major global mode of production, making up as much as 15 percent of world coffee production. Since 1988, other fair-trade marketing organizations (FTMOs) have begun to spin off into a well-coordinated international network[5] that works on the development of product and trading criteria, the development of a producer register for each targeted commodity, and global marketing of the message of fairer trade.

Coffee brands on sale in industrialized markets with fair-trade seals have between 2 and 5 percent of the market share, with the highest percentage penetration in Switzerland. Fairly traded brands have also achieved widespread distribution through conventional outlets (such as supermarkets). This illustrates that the fundamental premise of fair trade is seen by retailers as acceptable to the buying public.[6] This latter perspective suggests that the fair-trade experience in coffee may be pointing the way to a paradigm shift in the marketplace—something not revealed by a conventional statistical analysis of market share, turnover, and throughput.

What Distinguishes Fair Trade from Other Trade?

The many active parties in the fair-trade chain all play roles that are comparable to those in conventional trading—design and product development, production, logistics, transportation, marketing, customer service, and consumer mobilization. In fair trade, however, the motive, scale, orientation, and ownership, as well as the end messages, are distinctive.

In general terms, the differences are as follows. At the heart of fair trade is that the primary producers come first, not the product or even the consumer. Local skills, natural resources, and context are key. The alternative trader seeks to differentiate and to recast the value of inputs by the different parties in the trading chain, setting out deliberately to maximize the gains from trade that accrue to the Southern suppliers, the weakest links or protagonists in the trading chain, rather than to themselves or to superfluous intermediaries.

Fair trade therefore means in particular a good price to the primary producers. But notions of fairness are varied, as the attempts at conceptual differentiation listed below indicate. For the fair-trade movement, fairness can and does include in practice many elements apart from price. ATOs know the power and value of information about the market and seek to share this and thereby strengthen the bargaining power of their trading partners relative to other local and international markets. ATOs do not seek exclusive relationships with producers, an approach pursued by many commercial traders to consolidate the producer's dependency on them. ATOs recognize the need for product and organizational development and have provided relevant assistance wherever possible. Marketing has been reformed to mean, in particular, raising the awareness of the consumer to social justice and environmental aspects of trade in general, as well as of the brand or product being offered. ATOs frequently involve producers in this process, for example, by voicing their perspectives, concerns, and experiences as directly as possible to the consumer.

For a wide variety of other organizations that are not directly engaged in trading—political and labor organizations, social movements, environment and rights campaigners, and so on—fair trade has offered new opportunities for social organization and mobilization. A fairly traded product—whether a cotton shirt, an organic chocolate bar, or a CD—can offer a focal point for messages about how alternative economic theory might look in practice writ larger. Successful trading then becomes a tangible way to redress social and economic injustices of the past or to demonstrate the long-term viability of models based on more equal social relationships between buyer and seller, a more cooperative and less competitive exchange of goods between people. As such, many people have become attracted to the idea of fair trade because it provides both a strong statement about the way the world should be and a basis on which to critique the impact and effectiveness of the pre-vailing neo-liberal model. It is these latter effects in particular that provide the basis for real and lasting influence.

Does Fair Trade Work?

The fair-trade movement is not homogeneous. Some Northern trading com-panies in the movement are not-for-profit trading offshoots of NGOs work-ing in conventional development cooperation, such as Bridgehead (a division of Oxfam Canada). Others are now self-standing trading units, such as the Fair Trade Organisatie in the Netherlands, which also has a producer devel-opment arm, Fair Trade Assistance. Some have Christian roots and princi-ples: Traidcraft (U.K.), SERRV, and Selfhelp (U.S.), for example. Twin Trading (U.K.) was established by what was at that time London's regional government, the Greater London Council, and has a sister nonprofit charity, TWIN (Third World Information Network), that engages in research, publi-cation, and commodity trade development. Alter Trade in Japan and CTM in Italy have their origins and client networks in the consumer and worker cooperative movements.

The Southern trading partners in the movement are also varied. Most have some explicit community orientation, but with many different organizational forms and traditions. Some are cooperatives, such as the Tabora Beekeepers Cooperative in Tanzania; some are community-owned organizations, such as the trading associations of the Ñahñu in Mexico; some are parts of the gov-ernment, such as Tanzania's coffee parastatal, TANICA; others are based on traditional communal arrangements, such as the *ejidos* in Mexico, which date from the revolution; some are new forms of organization entirely, based on a mix of modern business and village or indigenous traditions, such as cocoa farmer organizations Kuapa Kokoo in Ghana and El Ceibo in Bolivia.

The heterogeneity of the movement makes it difficult to reach general conclusions of fair trade's effectiveness in offering a good deal to Southern producers. There is no single or commonly agreed on set of measurable cri-teria against which the quality of *all* trade can be assessed. However, there are some agreed-on criteria for particular products (such as MHF's criteria for coffee described earlier) and the broad Code of Practice subscribed to by all members of the International Federation for Alternative Trade (IFAT).[7]

Despite the efforts of IFAT and others, however, the effectiveness of fair-trade initiatives is hard to generalize through the prism of single and univer-sally applicable rules. Effectiveness assessments need, for example, to take into account the specifics of the market that is being entered and the supply-side realities and challenges. It is agreed throughout the movement, however, that a fair price is a critical element of fair trade. Indeed, as fair trade has entered the terrain of conventional trade and the consuming public, the

IFAT Code of Practice

1. To trade with concern for the social, economic, and environmental well-being of marginalized producers in developing countries—this means equitable commercial terms, fair wages, and fair prices, not maximizing profit at the producers' expense.

2. To share information openly, to enable both members and the public to assess social and financial effectiveness.

3. To reflect in their own organizational structures a commitment to justice, fair employment, public accountability, and progressive work practices.

4. To ensure a safe working environment, to provide the opportunity for all individuals to grow and reach their potential, and to ensure human conditions, appropriate materials and technologies, and good production and work practices.

5. To respect and promote development for and responsible to both people and the natural world.

6. To manage resources sustainably and to encourage production in a way that preserves and develops cultural identity.

Source: Based on IFAT. 1993. *Code of Practice.* Manila: IFAT.

argument that ATOs pay producers a fair price has become a major plank in the public positioning of fairly traded products. This focus reinforces the simple policy message of the FTMOs that the "'established' economic thinking fails adequately to describe production situations, where small-scale producers are forced to extreme levels of exploitation of themselves, and their families" (Transfair International 1995, 10).

What constitutes a fair price can be understood in many different ways. Listed below are some of the ways in which actors in the fair-trade movement have discussed and interpreted the meaning of a fair price:

- More than the local price currently available to the producer.

- More than the price available from other international traders.

- Enough for producers and their families to attain a reasonable or nationally recognized remunerative living standard.

- A price that enables the Northern partner to be no more than viable, but not to make a significant profit.

- A trading regime that allows Southern producers to earn the same as their Northern trading partners.

- Fixed remuneration to all parties involved directly in the chain, reflecting input, skills, and risk and not purchasing power or lending power alone— for mutual benefit.

A clear perspective as to what constitutes a fair price (and how it is arrived at) is needed by anyone wishing to assess the fairness of any particular trading relationship. For example, a British ATO, Traidcraft, has published in one of its social accounts a breakdown of the final retail price of several of the products it trades (Traidcraft 1994). It is clear from the examples in its accounts that the producer receives only a small proportion of the total price and that the British government receives more than the producer through value added tax. At the same time, Traidcraft indicated that it incurred a financial loss on this particular product. A reading of the full accounts for that year also indicates that the producers of this product got a higher price from Traidcraft than from local or other international commercial traders.

To complicate the matter further, the in-depth response to the question what is distinctive about fair trade transcends a debate that is only about price. All fair traders strive for a strategic and qualitative change in the trading process in favor of those who are traditionally the weakest. A floor price or other formula-based approaches give a strong message in a market climate that resists any and all controls or economic instruments (taxes, subsidies, incentives, and so forth), but at best it remains only a symbol or emblem of fair trade and its moral and economic objectives. An example of the breadth of issues that fair-trading institutions are considering is provided by the set of aims and principles for operating equitably between North and South (see the accompanying box), developed in 1985 at the founding conference of Twin Trading.[8]

Fair trade, then, aims to build better livelihoods for the poorest and weakest in the trading chain and to leverage developmental change and longer-term political shifts in their political and economic environment. Trade is the means to this end, and the signs of success might be manifested in a farmers' or artisans' organization as any or all of the following:

- Greater awareness, the ability to plan and think strategically, and the ability to participate proactively in the marketplace whether locally or internationally.

- Greater bargaining power through knowledge, experience, and access to infrastructure and inputs in the trading chain (via direct ownership or the ability to negotiate terms from a third party).

- Greater levels and equitable distribution of resources at the community level to invest in human capital for the future, the next generation.

Principles for the Development and Practice of Twin Trading

1. This statement of principles applies equally to South-South and North-North relations, as well as to those between North and South, although it must apply most particularly to the latter, since these principles have been most frequently disregarded in their case.

2. The basic principle of all exchanges between countries, whether of goods or services or technological equipment, patents, and know-how, should be in the mutual benefit of the peoples of those countries and not the profit of private capital.

3. All projects and exchanges should be directed toward abolishing inequalities of income and wealth and toward overcoming the results of existing unequal exchange and underdevelopment in the past.

4. Every effort should be made to prevent the establishment of monopolistic or monopsonistic positions in the market whereby prices of exports to the Third World are raised and prices of Third World products held down or rendered unstable.

5. All development projects, products, machines, tools, seeds, etc. and technology made available to the Third World should always be subject to the free and independent choice of the Third World parties.

6. Financial arrangements—credits, loans, grants, etc.—should be more easily available and freed of conditions, whether political or other, and should be designed to ameliorate rather than exacerbate Third World indebtedness.

7. All restrictions on trade imposed by First World governments, whether tariffs or nontariff barriers, especially on finished products, should be lifted so that Third World countries can enter a fair world division of labor.

8. Joint projects of First and Third World producers should be established to work out alternative technologies in appropriate fields to meet the real needs of their peoples and to conserve scarce energy and natural resources.

9. Third World producers should receive positive assistance to diversify their range of products and to increase their participation in the processing, refining, and marketing of their natural resources, possibly through joint agreements with First World producers.

(cont.)

10. Planned agreements for trade and technology exchange should be encouraged, including forms of barter and countertrade, not only between nation-state governments, but between regions, cities, and other localities, and between communities.

11. Training in the necessary skills for developing new technology and the opening up of professional posts to Third World candidates should be a major element wherever appropriate in development projects and technology agreements.

12. Corrupt practices, bribery, and backhanders of any sort must be eschewed and prevented in all trade relations and technology agreements.

13. Trade unions in First and Third World countries should seek opportunities for meeting together to draw up a code of labor for manufacturing industries in order to universalize best practices, such as the ILO Code.

14. Workers, whether in cooperatives or in other forms of economic organization, should be encouraged to develop their own decision-making arrangements and methods of work.

15. Equal opportunities for women, for all races and faiths, and for disabled persons should be guaranteed in all trade agreements and development projects.

16. The international boycott of trade and other economic relations with the apartheid regime of South Africa should be maintained and strengthened.

Source: Third World Trade and Technology Conference, statement of principles adopted at closing plenary, 22 February 1985, County Hall, London.

Fair Trade versus Free Trade:
The Experience of Latin America

The acceleration of grassroots efforts to organize has coincided with particularly adverse conditions in the last decade for farmers in the production and marketing of crops and commodities for both local and international markets. The promotion of alternatives—in the broadest sense of the word—has been a priority as well as a human and environmental survival strategy. In addition to historically low prices for many formally traded commodities (such as coffee and cocoa), the period has seen the reduction or withdrawal of government price supports, technical and other advisory programs, and virtually all state-sponsored protection for indigenous agriculture and basic food crops, notably corn (maize) and rice. This has occurred in the face of direct and often inequitable competition from imports. Changes in banking systems and an end to state support for agricultural credit have also meant a rise in exploitative loans and the consequent privatization of land via forfeitures. In other cases, land is being abandoned, with migration away from rural areas to the cities being the sole option.

How have farmers responded? What roles have external nongovernmental agencies and alternative traders played? What of the broader indicators of fair trade posited earlier? In this section, we present examples from Latin America and try to consider their impact objectively, including the farmers' own perspectives.

The Case of Coffee

In a recent public statement, an industry leader—Nestlé—claimed that its own activities in the market had a more beneficial effect on small coffee producers than fairly traded or, in its chosen language, "charity" coffees. The company stated that:

- The most important benefit to coffee growers is to sell as much coffee as possible.

- The price paid to growers depends on the balance between supply and demand.

- The international market is neutral.

- There are many ways of buying coffee, none of which is more or less ethical than another.

Nestlé concluded: "The charities are to be applauded for their efforts to help some cooperatives . . . but Nestlé's involvement in such work is already much more extensive."[9]

This public statement may ultimately have historical significance, in that it is the first time the company has entered into public debate with advocates and practitioners of fair trade implicitly *accepting* that a company's social and environmental footprints are a legitimate basis against which society can measure its performance. Of enormous importance, however, is to understand what such a statement means in the context of farmers' experiences of trying to survive in the international market for coffee, and the attempts fair-trade organizations have made to change this experience.

Important, in this context, is to fully understand the changes that have been occurring in the last decade in the coffee regions of Latin America and to the people who work and live there. It is therefore necessary to review what happened in the coffee market in the period 1989 to 1995. Without this backdrop and some analysis of the causes and effects of this boom and bust, the scale and scope of the efforts made by farmers to organize and the extent to which fair-trade coffee defied the market rules are hard to appreciate.

Failure of the International Coffee Agreement (ICA). The principle cause of the price crash in 1989 was the failure of the coffee producers and consumers to renegotiate an agreement to take effect on expiry of the prevailing one. Consumers, meaning consuming *countries*, and particularly the most powerful market makers in the United States, no longer wanted to support an agreement regulating the supply from coffee producing countries because:

■ The agreement did not enable the United States to purchase what it wanted, which was less robusta coffee and more mild washed arabicas—notably from the producing countries of Central America. These factors reflected gradual but irrevocable changes in coffee-drinking preferences and hemispheric geopolitics.

■ They wanted to stop the cheap (non-ICA controlled) coffee going into the Soviet and eastern European markets from being resold to the West illicitly (so-called leakage).

Consequently, the International Coffee Organization failed to broker a new agreement that hinged on a change in the allocation of "quota"—the quantity of bags of coffee a producing country may sell in a given year—to adapt to a new situation and pent-up market demand. Only Brazil had surplus quota and would not concede any part of its share. No agreement could be reached on how one producing country could be allocated a quota. The historic but profoundly inappropriate methodology was based on volume of production. The result was the liberalization or freeing of a formerly controlled or at least arbitrated market, with great volatility in prices and production the result. Central American producers were early winners in a narrow sense. The losers were the mainly African robusta and other lower-grade coffee producers. Nobody directly involved in growing, processing, and marketing coffee benefits from the uncertainty of price and supply of quality coffee.

"Short Termism" and the Loss of Relationship. The 1989 bust revealed to a new generation of coffee traders a simple truth: the longer you wait, the cheaper the product. Earlier coffee trading practices reflecting "buffer mentalities" and gentler times—holding stock, stock financing, and longer lead times—were cast aside. Just-in-time practices reached the coffee trade, with the added incentives for brave or well-connected traders able to drum up decent coffee at the last minute of cents-per-pound savings or extra margin by the minute as prices descended.

Facts and Figures

1989 International Coffee Agreement (ICA) reaches its renewal date with nothing in place to regulate supply and demand. Stockpiled coffee in consuming countries' warehouses is at 20 million bags plus—higher than traditional average levels (two- to three-month supply).

1989 Price crashes. Bust sets in.

1992 Price reaches 50 cents a pound—1930s levels. Producing countries meet and aim for a coffee retention plan.

1993 Coffee retention plan gains wider support, including essential participation of Central American and verbal support by Brazil.

1994 In spring, the market waits for the new Central American crop. The retention plan kicks in—price rises from 60 to $1.40 a pound.

1994 Stocks held in consumer countries are beginning to dwindle. Frost hits once: price rises to $1.80.

1995 Frost hits a second time: price rises to $2.73—a mini-peak. Price falls back to $2.40, $1.80, and reaches 90 cents at the end of 1995.

1996 Price rallies to 90 cents to $1.30 range. Consuming-country stocks are low. Except for Brazil, the producing countries have no stocks. Market remains sensitive and very volatile.

Sources: Twin Trading records and coffee trading staff.

Booms and Busts—The Risks and Benefits of Financing Coffee. The levels of stocks of green coffee in consuming-country warehouses are a critical factor. In the bust and boom cycle described in the 1989–96 period, stocks were held at higher than normal levels initially. This fallback stock checks any rise in supplier prices.

The facts and figures box highlights the relationship between declining levels of stock in consuming countries, rising levels in producing countries (via the retention plan), and recuperation of the price. For most suppliers of green coffee, the price rose in a clearly orchestrated manner, the harvest period, and gains fell principally to importers and end users. Few small-scale producer organizations have recourse to the financial resources to hold on to their product; for lack of liquidity, they are forced to sell to the local speculator or intermediary (*acaparador*). Farmers lose ownership of their product relatively fast. They are seldom able to consciously enter into this macro-level market-shifting dynamic and attempt to influence it to their advantage or at the very least to minimize risk.

Similarly, few producing countries can sustain the levels of liquidity or the political cohesion necessary to plan and play a permanent buffer stock role. The producing-country retention plan attempted during this period can be said to have worked—given the market demand for mild coffees, the solidarity and cooperation of Central American and Colombian producers were pivotal—and the initiative and collaboration shifted the market. But few commentators are certain that Brazil, a major player in every regard (scale of production, historic significance as a producing country, and so forth), ever actually held and financed new crop stock in accordance with the scheme.

Coffee Production and Immigration to the Cities.   Since the late 1980s, the price of coffee is generally thought to have fallen well below the viable cost of production. Break-even is assumed to be 70 to 90 cents (US) per pound, depending on the type and mode of production. At this level or lower, costs remain uncovered, and labor is underrewarded; plant care declines, and with it quality and yield; income levels of families fall, with other socially and economically inevitable results: impoverishment, malnutrition, migration from the rural areas, and great damage for future generations—human capital for these marginal communities.

1989–95: The Case of Mexican Coffee[10]

As the market tightened about them, farmers from all over Mexico came together in a process of definition, solidarity, and reorganization. This was not an isolated process, and the efforts spread over state and national borders and across continents. The following local-to-global summary shows the depth and range of actions initiated and undertaken in the tense and difficult period under review.

Locally.   Longtime organizational struggles for land and for direct dialogue with the state and governors; with the national institution for coffee, INMECAFE; with the state and private banks; and with state credit and technical and agricultural service institutions came to a head.

State level.   Many farmers' organizations were reestablished and refocused into new and more independent unions of cooperatives, *ejidos*, or state-level networks, such as the regional coffee producer organization in southern Mexico, CEPCO, or were given access to external credit from ethical banks.[11] Other nongovernmental institutions supported the efforts to meet, diagnose, and prepare strategies, for example, Oxfam and the Inter-American Foundation.

Nationally.   The national coordinating body of independent coffee producers, CNOC, was formed by about 65,000 coffee farmers organized into local or regional unions, federations, and cooperatives in six states. Its own

Price Speculation and Volatile Markets: Views from the Farm

"For four years, no, longer actually, since 1986, we have been really badly off. Now the new [rising] prices finally compensate for this. They give us something to work with. Finally there should be some money to invest in technology. We've been doing everything the hard way . . . by hand. Lots of members in the co-op didn't give up the struggle, didn't give up tending their farms and being good coffee farmers, though."

Q: How are you dealing with this?

"Each co-op has an education committee—apart from technical training, they are responsible for keeping members informed about the markets [and] clients through meetings and discussions. Our co-op is a chapter of OCIA [Organic Crop Improvement Association], the certifying body, and as such we were able to sell our organic coffee at around US$103 per quintal. This is at a time when local traders were paying 100 soles [around US$50]. So many members are aware [of this situation] and are happy. Some are not [aware] and they will feel let down if we cannot compete now [that] the traders are finally back."

Source: Conversation between Pauline Tiffen and Mario Villalobos Perez, president of Bagua Grande coffee co-operative, northeast Peru, 1994.

export trading arm, the Promotora de Cafes Suaves de Mexico, was set up, and five CNOC member organizations invested in and supported the development of a company to own and promote a new brand for their coffee in the United States—Aztec Harvest.

Regionally. The small- and medium-scale coffee producers' union UPROCAFE, the Latin American coffee producer Frente de Cafetaleros Solidarios de America Latina, the Via Campesina in Central America, and the South-South Network (Enlace Sur-Sur) in the Caribbean were formed to exchange experience and offer support.

Intercontinentally. The Small Farmers Cooperative Society (SFCS), including coffee farmers' cooperative unions from East Africa, was formed in 1992. Market-based TWINcafe provided information, New York futures hedging and options services, a credit fund, and market representation for SFCS members. The First International Conference on Organic Coffee was held in Chiapas in 1994, with 200 international visitors and participants.

Mexican Farmers Access the Market[12]

From small beginnings, another process also began in this period. Only three Mexican coffee organizations had regular fair-trade sales opportunities

---

Social Consequences of Coffee Market Busts and Booms from the Perspective of Coffee-Growing Communities, Northeastern Peru

"Coffee is *the* cash crop here. Bad prices mean great hardship. Continued bad prices (as we have seen since 1986) influence all activities in the community . . . In school we really see the effects in our students. They are eating less and so their attention span is poor, they take poor notes, they are tired and can't concentrate. Some stop coming completely: the walk is too far, or if they are sick, their parents can't afford medicines. The parents can't afford to buy books or uniforms.

"When there is no work, people, especially the younger, more educated ones, leave. . . . they go and work on 'the other crop' [coca].

"In times of low [coffee] prices malnutrition stands out—skinny children, ribs showing, no shoes and a variety of diseases . . . for example walking to school with no shoes makes kids prone to getting *pique* [a parasite that bores a hole in the sole of the foot]."

—Carlos Carranza Fuentes, teacher at the school for the last five years, in charge of the school medicinal herb garden

Source: Interviews by Pauline Tiffen, 1994.

---

before 1993—the Union of Indigenous Peasants in the Isthmus Region, UCIRI (Oaxaca); the Atoyac *Ejido* Coalition, *Coalición de Ejidos* (Guerrero); and the coffee producers' union of the southern region, UNCAFESUR (Chiapas). Therefore, greater pressure was felt to develop a *comercializadora* or trading arm oriented toward, and competent to deal within, the conventional coffee market. Furthermore, Mexican coffee attracts a duty on entry to the European market, as it is not of African, Caribbean, and Pacific country (ACP) origin; it gets no special treatment under the tariff arrangements of the General System of Preferences and is excluded from the exemptions afforded Andean coffee producers under various antinarcotics programs. The tariff has acted as a disincentive to export Mexican coffees into Europe and has accentuated the need to develop strategies for better access to the U.S. market.

In 1988, the coffee producer organization Coalición de Ejidos de Atoyac exported 250 bags of green coffee without passing through the national peasant farmers' confederation dominated by the ruling party, the Confederación Nacional Campesina (CNC). At the same time, the parastatal coffee marketing institute, INMECAFE, was successfully lobbied to give quota stamps directly to a farmers' organization for the first time. By 1994–95:

- Second-level farmers' organizations had reached levels of organization sufficient to achieve trading norms and were exporting via the Promotora to export markets.

- The grassroots constituents of the eleven primary-level coffee producer organizations numbered at least sixty, with an estimated 45,000 farmer members.

- All eleven organizations had made contact with and broken into the alternative or fair-trade market, adding to the returns and sense of contact and achievement.

- Aztec Harvest—the direct promotional sales initiative set up in 1992 to access the U.S. gourmet and specialist organic coffee market—grew to incorporate coffee from nearly 1,000 farmers in forty-two villages.

- During the 1994 harvest period, the Promotora de Cafes Suaves de Mexico traded 51,354 bags of green coffee to U.S. and European markets with an approximate value of US$6.5 million FOB.

- The Unión de Ejidos de la Selva (Chiapas) opened a coffee shop in a fashionable part of Mexico City to promote the consumption of high-quality coffee and increase the Mexican public's awareness of environmental and social issues in Chiapas.

Gaining the premiums on fair-trade contracts at times when prices were below the cost of viable production was undoubtedly crucial and made the difference between bankruptcy and survival for many small-scale coffee farmer organizations in this period.

However, the add-on effects of access to fair-trade markets and increased activity by farmers in other trading arenas may be far wider and of greater significance in the longer term. The effects are both quantitative and qualitative. In Tanzania, for example, a cooperative union can enter the formal auction where, under the current national system, all exporters must meet to buy coffee and can bid up the price of all the co-op's coffee to other buyers. With Cafedirect or other fair-trade forward contracts in hand, the other buyers are being forced to compete in a way that would not otherwise happen. Fair trading alongside conventional trading is, in a small way, ratcheting up other companies' standards of behavior, producer expectations, and product prices to increase return for farmers. The value of this add-on benefit has not been calculated.

To conclude this coffee case study with a short qualitative assessment, it seems appropriate to use the views of a coffee farmer.

A few years ago, before initiatives like Cafedirect and Max Havelaar, it seemed as if farmers were only good for walking long distances with coffee on their backs. . . . I think that the *comerciantes* (local traders) just couldn't imagine

that one day we'd be working directly with people from outside. Not just that
. . . but that they'd be interested in us enough to come and visit. The *comerciantes* certainly never thought we'd learn to trade ourselves.[13]

The Bigger Picture

Environmental Dimensions of Fair Trade

> Unless we build markets for forest products and make sure the benefits
> flow back to local communities, forests will continue to fall to loggers,
> ranchers and farmers who see no other option.
> —Mike Saxenian, Conservation Enterprises director, 1991

Given the constraints imposed by the economic system and consumer interest in low-priced products, achieving sustainable agriculture and environmentally neutral trade is a challenge. One particularly problematic aspect is the common but artificial distinction between environmental measures and impacts, on the one hand, and social and economic policies or regulations, on the other. This division of the "indivisible" is often the bane of policy process and debate within and between international organizations.

The following gives some idea of these issues as they affect agricultural production and fair trade:

■ *Technical:* (1) Integrated pest control, whether with minimal or zero chemical inputs, may reduce yields and thus producer earnings. Who pays for this is a question that the free market is unable, and governments are unwilling, to address. (2) There is frequently a local imperative not to address even clearly unsustainable practices that are detrimental to the soil and fertility.

■ *Colonial attitudes:* The failure of Northern consumers to adjust their consumption practices and expectations, while at the same time attempting to regulate against the entry into their markets of so-called unsafe or unsustainable products, is a political issue and suggests that we may be entering a new era of biocolonialism.

■ *Ecological footprints:* Trade across long distances is seen as contrary to the principles of sustainable practice, because of the use of nonrenewable fossil fuels and the resultant displacement of local production and thus agricultural self-sufficiency.

■ *Human rights:* Human rights are not guaranteed through organic production. Although there is clear complementarity between environment-friendly and people-friendly agriculture, there is not an absolute overlap. For example, certified organic production can be undertaken in exploitative conditions and with poor or nonexistent labor provisions (and vice versa).

The debate about the environmental dimensions of fair trade and international trade in general still lacks an appropriate framework to assess what might be termed "competing negatives" in an equitable way. In the short term, fair trade aims to offer a response to people facing oppressive poverty and to increase their ability to defend their rights. However, there is a need to explore more fully the linkages between fair trade and long term sustainable development imperatives (see Zadek and Tiffen 1996, 1997).

Socially Responsible Business

Another area of fair-trade activity has been the growing support within the broader community of socially responsible businesses. These include a wide range of business sectors, including banking. Most prominent among the retail and brand-based chains have been the British cosmetics company The Body Shop and the U.S. ice cream company Ben & Jerry's Homemade. Both have experimented in high-profile fair-trade initiatives as part of their overall socially oriented policies and practices. The Body Shop has traded with communities in several Latin American countries, notably Brazil and Mexico. Ben & Jerry's fair-trade purchases from Latin America have included coffee from Aztec Harvest, which is used to flavor the company's ice cream products.

Unlike the experience of the ATOs to date, the claims of commercial companies to be trading fairly with small producers have been increasingly challenged. Companies like Ben & Jerry's and The Body Shop have been accused through the media of exaggerating and distorting their activities for public-relations purposes. Some companies have sought to open their business practices to external inspection, partly in response to such criticisms. The Body Shop and Ben & Jerry's in particular have moved toward comprehensive social auditing of their activities (Zadek and Raynard 1995).[14] The approach of these companies to social auditing has included discussions and consultations with a range of stakeholders in the company (including Southern producers); the selection and application of key benchmarks, indicators, and targets; external verification; and public disclosure. For example, The Body Shop's recently published *Social Statement* included independent assessments of its "Trade Not Aid" activities with Southern producers. The document highlighted a mixed record of achievements, but with an overall picture of positive economic gains to Southern producers (Body Shop 1996).

Interestingly, both of these companies involved proactively in fair trade have interpreted the principle behind fair trade as one that they wish to apply more widely, that is, to include relationships with community-based producers in the North as well as in the South. This is a noteworthy and legitimate addition to the alternative trader's original North-South concept, which properly belonged to a period of postcolonial independence and pre-dates the reality of the global market. The trading principles of Twin Trading, noted earlier, for example, show a distinct North-South division of labor and North-South aspect to the problem of unequal bargaining power between parties in the economic system. They could be reworded to reflect the shift in axis to a new economic divide and the globalization of need and economic difficulty for people working at the margins in all societies.

The Marketplace and Consumers

There are a number of factors influencing the evolution of fair trade. Undoubtedly, however, retaining the interest and effective engagement of consumers and their commitment to purchase fairly traded products over the unfair variety is a major challenge. A consistently positive view emerges from surveys of consumers' ethical concerns—the majority of whom say that they support the notions behind the fair-trade concept.[15] However, this concern has not yet translated into purchases. Furthermore, according to the director of Transfair International, Martin Kunz, consumers cannot easily distinguish between false and genuine claims, even with the rise in the presence and authenticity of seals and marks denoting more carefully substantiated claims (chemical free, fairly traded, recycled contents, and so forth). Other European fair-trade initiatives involving seals of approval have experienced a constant and uneven battle between the pulling power of their own endorsement compared with the credentials of a familiar, heavily promoted mainstream brand or a corporation perceived as solid and trusted.[16]

There are two phenomena, however, in the domain of the consumer that are provoking companies to review their own operational systems: the capture of market share by the new fair-trade brands, and the negative impact on sales resulting from public response to disclosures of poor or negligent corporate performance in the social and environmental areas.

Fair Trade as Catalyst—Qualitative Competition

The other driving force behind the success of the fair-trade movement and the take-up of social issues by mainstream corporations is its capacity to influence mainstream business directly by its very presence in the market. The scope of this phenomenon may just be emerging—namely, the ability to transfer the operational principles of fair trade, without compromise, into extremely effective and competitive business. Fair trade—trade that systematically weights its total quality and operational systems in favor of primary producers and is rewriting the rules on consumer communication—has begun, in practice and in the full face of competition with conventional private companies, to prove effective and successful at several levels, including across normal commercial indicators. In this way, fair trade is beginning to achieve a crossover into the territory of conventional business: taking market share in the industrialized countries, and outperforming conventional companies and their trading systems in the Third World context.

Three examples illustrate this critical approach to achieving leverage over markets dominated by large multinationals.

The El Ceibo Cooperative. This cooperative in the Rio Beni region of Bolivia was set up in 1978 by migrants from the highlands who originally arrived in the region in the 1950s. By 1994, the cooperative's annual earnings from sales of organically certified cocoa beans and chocolate sold to health-food stores and fair-trade markets and outlets reached US$600,000. El Ceibo—a federation of thirty-seven cooperatives with about 900 farmer members and 100 employees—has attracted attention for its integrated approach, which has built its international commercial activity onto a social and organizational system based on traditional practices, including equitable distribution of earnings, consensus-building assemblies, and respect for reciprocal obligations. The marketing has effectively targeted niche and specialist segments of the market—namely, the ethical and organic sectors (Thrupp 1995, 124–25).

Kuapa Kokoo. This is the only farmer-owned and -run company founded in Ghana since the liberalization of the internal market for cocoa in 1993. It was set up by and is run with the assistance of Twin Trading. After three years in operation, with around 5,000 farmers and trading 4,000 tonnes of export-grade cocoa per harvest, it has overcome much of the early prejudice it experienced as a farmer-oriented operation and is recognized as the highest performing licensed cocoa trading company in Ghana (except for volume). Kuapa Kokoo is profitable, is paying healthy bonuses to members, and delivers excellent quality cocoa. It has an effective operational system based on just-in-time and village-level cooperation, operational control, and trust, with a rapid turnaround and use of capital (this is the largest business cost in Ghana, where interest rates reach 45 percent). Its loan repayment record is second to none in the sector. Other cocoa producers are gradually being obliged to reorganize and to upgrade their treatment of farmers in the rural areas.[17]

Cafedirect. This is the United Kingdom's leading fair-traded coffee brand, now available in most retail outlets throughout the country. Cafedirect not only guarantees to its suppliers that it will never follow the market down below a floor-price level ($1.26 per pound, based on earlier International Coffee Organization assessments and the Max Havelaar baseline criteria) but also pays a standard 10 percent premium on top of the market price. In mid-1992, with green coffee prices at 50 cents a pound, Cafedirect did not flinch from its bottom-line policy, and consumers continued to support the newcomer on the shelves. Prior to the launch of

Cafedirect in 1991, conventional market wisdom indicated that there was little, if any, market share to be obtained by any other method than purchasing it, which means the merger or acquisition (and revamping) of an existing brand. In the performance-conscious environment of the mainstream market for fast-moving consumer goods, Cafedirect, now with a 3 percent market share in the smaller but significant roast and ground market, has challenged the common wisdom, quite possibly irrevocably. In particular, this has been done by offering a product of impeccable quality with a clear raison d'être way beyond the normal product selling propositions. Sales of the product and market research both indicate that consumers are not immune to non-conventional product-related propositions and that Cafedirect may have precipitated a tangible crossover in purchasing interests—from the often registered ethical intent to the actual purchase, made in a conscious and reasoned way.

What all the fair-trade initiatives have shown—with their capture of between 2 and 5 percent of market share—is that even minor shifts in consumer purchasing power register high on the "Richter scale" of corporate management priorities and things to watch. There is absolutely no reason to believe that this trend—and thus mainstream companies' interest in addressing the threat or the opportunity of fair-trade markets—will diminish as long as nongovernmental and campaign organizations continue their efforts to reach and articulate issues of concern to consumers—whether the human and environmental costs of farming or clean clothes—and as long as consumers continue to find them credible.[18]

Corporations risk a general lessening of their credibility because of commercial and industrial accidents and errors of communication, and consumers have experienced directly and indirectly greater economic uncertainty at the hands of companies engaged in downsizing and streamlining exercises. Fair trade—as a complementary variant on the environmental cradle-to-grave or life-cycle approach—is increasingly positioning itself as one of the few solutions to the apparently profound corruption in the trading chains, where profit is put before people at all levels. It is interesting to notice the audible and strong echoes of the FTMO concept in the following May 1996 statement from a major U.K. retailing chain:

> We buy our fresh beef from a *strictly limited number of approved suppliers* in the UK and Ireland. We *work in partnership* with these suppliers and the farmers to ensure the safety and quality of [our] beef and are therefore *able to keep track of all stages and links* in the production chain."[19]

However, so-called mad-cow disease—the subject of this leaflet and the reason for a reassuring consumer message—is only the latest in a long line of trading and production system failures, with a direct and adverse impact on people: non-tamper-proof Tylenol, HIV-contaminated blood products, poorly supervised building regulations, and so forth.

At any scale, and for most economic sectors, it is necessary to question whether legislation is part of the equation of future success in fair trade. The World Trade Organization rules, for example, oppose regulation that judges the quality of a product by the way in which it has been produced. The major thrust of opposition to this approach has been premised on its environmental implications. However, the social justice dimensions of the approach are equally open to criticism, since WTO rules currently forbid countries from banning the imports of goods that, for example, are produced using labor in highly exploitative conditions. Similarly, it is not impossible to see a place for the operational norms and principles of fair trade as an integral part of conventional business quality standards and quality management practices, such as those set by the International Standards Organization. Fair-trade practice and initiatives are feeding directly into this wider lobbying process.

> Alternative trade can be viewed as a flea on the hide of an elephant, a tiny presence capable of making its giant host sit up and take notice. . . . The alternative trade movement, by grappling with what conditions should make up "fair trade labels" can help movements of workers, environmentalists and others trying to define new labour and environmental parameters in trade agreements. (Barnet and Cavanagh 1994, quoted in *Jakarta Post*, December 1994)

"Fair trade" is a description of an experience and of multiple interlocking efforts by organizations that produce, trade, and distribute and that seek to promote and shape the way trade is done in practice. These organizations seek to achieve their aims by tackling a range of both traditional and non-traditional commodities, constantly developing new insights into sustainable production and marketing, and in this way challenging the ideology of free trade, where costs and benefits are narrowly defined and seldom include the wider issues of welfare, mutual benefit, and the common good.

Success beyond the small scale of the current movement will not arise only by adding more fairly traded products to the shopping baskets of consumers around the globe, that is, by simply increasing market share in the traditional sense. Rather, a growing market share will mean that the mainstream market players will have to respond increasingly to the demonstration that this trade represents and to the interests of consumers in buying on the basis of broader social and environmental criteria. The original alternative traders have an important role as the "compass"—the "true north" for the movement as it expands.

Larger-scale success will depend on other developments, including greater synergy among the various strands of the fair-trade movement and the continuing and honorable uptake of the method and goals by mainstream corporate business. Quantitative change is, however, required for these forms of qualitative trading and innovation to be converted into effective challenges to the dominant inequitable and environmentally destructive trading patterns.

Notes

Thanks to the many people who gave the authors useful comments on this chapter. Thanks in particular to Michael Barratt Brown, Bert Beckman, Jutta Blauert, Jacqui McDonald, Carol Wills, and Rachel Wilshaw.

1. Other useful documents include *Fair Trade: Guide to Good Practice*, Towns & Development, 1995, *Bridge Framework Manual: An Explanation of the Fair Trade Programme of Oxfam UK and Ireland* (February 1996), and *Fair Trade: A Rough Guide for Business* (Twin Trading, 1994; reprinted, 1996).
2. For example, the Environmental Economics Department of UNEP now openly notes and rejects as inadequate otherwise classic economic indicators such as gross national product for the tracking of environmental developments (ICO Conference on Coffee and the Environment 27–28 May 1996).
3. The earliest ATOs in their modern form were Oxfam Trading (then Bridge) in the United Kingdom and Fair Trade Organisatie (or its predecessor) in the Netherlands.
4. The foundation is named after the hero of a Dutch nineteenth-century literary classic—a work of fiction in the mold of *Robinson Crusoe* that features a young man's realization of the inevitable conditions for workers on the coffee plantations in the Dutch colonies of Southeast Asia.
5. Including TransFair International in Germany, Austria, and the United States and Canada; the Fair Trade Foundation in the United Kingdom; and Max Havelaar in Denmark, Sweden, and Switzerland.
6. Gaining access to supermarket distribution is significant, because the trend is away from local shops to large-scale retailing. In the United Kingdom, four retailing companies (supermarkets) take almost 50 pence out of every £1 spent on the purchase of groceries (Cowe 1996).
7. IFAT was founded in 1989 and aims to provide a forum for Southern producers and Northern importers.
8. Delegates to the conference came from NGOs and from progressive local authorities in Britain, continental Europe, Africa, Asia, Latin America, and the Caribbean at the invitation of the Greater London Council to discuss and to conceptualize a new approach to trade and exchange between people of the North and the South and thereby to contribute to the initiation and launch of Twin Trading and the associated registered charity TWIN.
9. From a consumer fact sheet issued by Nestlé U.K. in June 1995 entitled "Nestlé Does More Good for Coffee Growers in the Third World Than All the Charity Coffees Put Together."
10. Sources for this section include CNOC 1991; Twin Trading records; and David Griswold, "In Harmony with the Rainforest," cited in Thrupp 1995.

11. The Ecumenical Development Cooperative Society, headquartered in the Netherlands but with project and loan officers situated throughout Latin America, is an ethical bank that since the mid-1980s has provided substantive support to supplier organizations connected to fair-trade initiatives, as well as directly to the Northern counterparts. Other such banks include Shared Interest and Mercury Provident (now Triodos), United Kingdom; Triodos and Hivos/Stichting Doen, the Netherlands; and RAFAD/IGS, Switzerland. The latter are part of a global network of ethical, environmental, and anthroposophical banks supporting fair trade.

12. Sources for this section were Twin Trading records; Jos Algra interview; Novotrade Consult b.v., the Netherlands, records; Ransom 1995a.

13. From Pauline Tiffen's 1994 interviews with Peruvian coffee producers.

14. Ben & Jerry's has been carrying out social assessments for eight years, and the first Body Shop social audit published in 1996 covered ten stakeholder groups, one of which was fair-trade suppliers.

15. For example, in a recent survey by the Cooperative Wholesale Service, 60 percent of those surveyed said that they would be willing to boycott a shop or product because of concern about environmental or social performance. Almost 80 percent then said that they felt that retailers and manufacturers should provide more information on environmental matters and ethical issues in order to help them make their purchases.

16. Based on a presentation by Martin Kunz of TransFair International at IFAT Regional Conference, Oxford, May 1996; positions of Transfair International (1995); and correspondence with the former general secretary of Max Havelaar, May 1996.

17. Based on Max Havelaar, International Cocoa Register reports; Heini Conrad; TWIN records.

18. For information on specific product or sector trade campaign reports, contact the following U.K.-based organizations: World Development Movement, Oxfam, Christian Aid.

19. Safeway consumer leaflet, "A Guide to the Facts about Our Own Brand Beef Products" (emphasis added).

References

Barnet, Richard J., and John Cavanagh. 1994. *Global Dreams: Imperial Corporations and the New World Order*. New York: Simon & Schuster.

Barratt Brown, M. 1993. *Fair Trade*. London: Zed Books.

Body Shop. 1996. *The Body Shop Social Statement 1995*. Littlehampton: Body Shop.

Bridge. 1996. *Bridge Framework Manual*. Oxford: Oxfam.

CNOC. 1991. *Cafetaleros. La Construcción de la Autonomía*. Cuadernos Desarrollo de Base no. 3, Mexico City.

Cowe, Roger. 1996. Supermarket Forces Offer Mixed Blessings. *The Guardian*, 13 April.

IFAT. 1993. *Code of Practice*. Manila: IFAT.

MacGillivray, Alex, and Simon Zadek. 1995. *Accounting for Change: Indicators for Sustainable Development*. London: New Economics Foundation.

The Mystery of Growth. 1996. *The Economist*, 25 May, p. 16.

Niessner, Birgit. 1996. Looking Behind Fair Trade. Dissertation, London School of Economics.

Ransom, David. 1995a. Coffee: The View from the Valley. *New Internationalist* 271 (September).

Ransom, David. 1995b. The Well-Informed Coffee Drinker's Directory. *New Internationalist* 271 (September).

Thrupp, Lori Ann. 1995. *Bittersweet Harvests for Global Supermarkets: Challenges in Latin America's Agricultural Export Boom.* Washington, D.C.: World Resources Institute.

Traidcraft. 1994. *Traidcraft Social Audit 1993/94.* Gateshead: Traidcraft.

Transfair International. 1995. Some Basic Thoughts on the Future of Fair Trade Marking. Working Document 12.95, Version 1.2.

Zadek, S., and P. Raynard. 1995. Accounting for Change: The Practice of Social Auditing. *Accounting Forum* 19 (Autumn): 164–75.

Zadek, S., and P. Tiffen. 1996. Fair Trade: Business or Campaign. *Development* 3 (Autumn): 48–53.

Zadek, S., and P. Tiffen. 1997. Fair Trade: Paths to Sustainability? Working paper, prepared as part of a submission to the United Nations Commission for Sustainable Development, New Economics Foundation, London.

Part IV

Organizing for Mediation

7

Building Sustainable Farmer Forestry in Mexico

Campesino Organizations and Their Advisers

Gerardo Alatorre and Eckart Boege

During the 1990s, neo-liberal policies have had several effects on the eco-nomic initiatives of the Mexican peasants. On the one hand, several peasants' enterprises are in crisis due to market problems, lack of credit, a general drop in public social expenditure, and adverse legislation (de Ita Rubio 1994). On the other hand, the withdrawal of the state from economic life has opened up new opportunities. Some groups have been able to take advantage of market deregulation in order to strengthen their organizations and build up their market positions. However, the ability of peasant groups to face new economic and trade policies has depended on a wide range of issues, most notably their productive strength, their alliances, their internal organizational patterns, and their leadership capacity.

In this chapter, we draw on two concrete experiences as the basis for a more general analysis of the involvement of peasants and the leverage exerted by mediators—the peasant farmers' advisers. The experiences dealt with here are (1) that of the communities and *ejidos* (farmer communities on land granted by government decree) in southeastern Mexico and the processes they followed in order to take on the management of the rain forests they inhabit, and (2) that of several nongovernmental organizations (NGOs) that, after an extended participation in the emergence and develop-ment of peasant forestry enterprises, are coordinating the efforts made by these small businesses.

Forests in the Hands of Peasants

In Mexico, *ejidos* and rural communities own about 80 percent of the sur-face covered by forests.[1] Nearly 17 million people inhabit the forested regions, depending for their livelihoods mostly on these same forests (Chapela 1992, 342). There are 21 million hectares with logging potential, but only one-third of this surface has government authorization for its exploitation. Of the 29,983 Mexican *ejidos*, about 7,000 have forest areas, with just over half of these owning exploitable forest areas (González Pacheco 1992).

Forestry production in the country faces a severe crisis, reflected by the drop in production from 9 million to 6 million cubic meters between 1989 and 1993. Moreover, more than 50 percent of the infrastructure of the industrial logging sector is obsolete. The accumulated ecological liabilities in this losing game for nature and peasant families have to be attributed to this obsolete system, and not to the peasant producers, who are condemned to sell their raw material below value. The paradox is that even though peasants and the indigenous communities are the owners of nearly 80 percent of the forest surface, their production amounts to only 17 percent of the total (*La Jornada*, 1 December 1995).

Despite the fact that most forest areas are still covered by communal land-use laws and access regulations, over many decades, exploration rights to forests have been granted to private companies. Peasants received from these enterprises only the *derecho de monte* (a form of land rent), which was only a small fraction of the commercial value of the timber. In the state of Quintana Roo, for example, the government-affiliated enterprise Miqro has extracted 600,000 cubic meters of mahogany from some 400,000 hectares of *ejido* land over twenty-five years. Another company located in Zohlaguna (state of Campeche) practiced nonselective deforestation[2] on 400,000 hectares of national and *ejido* lands.

Yet a transition toward peasant ownership took place in several parts of the country during the 1970s and the early 1980s. This transition was the result of a series of peasant mobilizations,[3] a favorable political context, and the termination of several government concessions to large logging enterprises. In Zohlaguna, for instance, the peasants undertook a ten-year fight against the company for better prices and for control of the exploitation. During the same period, a change was also notable in the peasant movement at the national level (Bartra et al. 1991; Chapela 1991). After the boom in the agrarian struggles of the 1970s, several organizations shifted their efforts toward consolidating their social enterprises. At the same time, a gradual strengthening of regional and national organizations took place.[4]

In the case of peasant forestry enterprises, several technicians from governmental institutions and NGOs played an active role in the technical and administrative training of the peasants, who effectively turned into organized forestry producers. This process was observed in the Yucatán Peninsula (states of Quintana Roo and Campeche), in Huayacocotla and the Cofre de Perote (both in the state of Veracruz), in the Sierra Norte of Puebla state, and in some regions of the states of Michoacán and Oaxaca.[5]

In this chapter, we describe the experience of the peasant organizations in the Yucatán Peninsula. In the peninsula, 3 million hectares of the states of

Figure 7.1   Forested Areas and Biosphere Reserves in the Mayan Zone (Mexico-Belize-Guatemala Triangle)

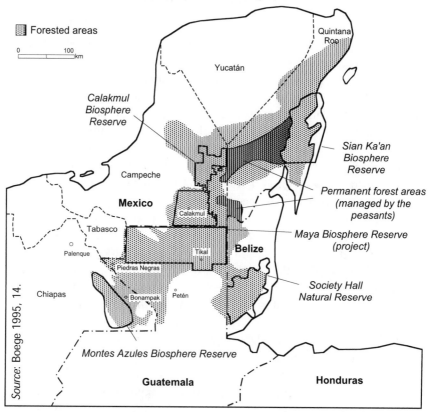

Quintana Roo and Campeche are covered with forests, an area adjacent to the Guatemalan Petén and the forests of Belize (see Figure 7.1). Several regional forest producer organizations exist today in over eighty *ejidos*, with a total area of 500,000 hectares of permanent forest areas and a large proportion of indigenous population.

In the communal and *ejido* lands, experience has been gained by both local and regional organizations in developing a sustainable and economically viable forestry production system. This experience highlights the need for success in the support of what is still a scarce collaborative effort. In this case, effective interinstitutional collaboration was achieved through the assistance of several public institutions and a new technical team noted for being independent of the otherwise corrupt staff of the Forestry Department, and through the peasant organizations interested in defending their forest resources. The most relevant feature of this experience, as compared with similar cases, is the positive conjuncture of different actors collaborating at the local, intermediate, and national levels.

New Opportunities for Change:
The Public Sector

In 1976, Cuauhtémoc Cárdenas, one of Mexico's current opposition leaders, took office as undersecretary for forestry and fauna, promoting a scheme of peasant forestry. Amidst a fight between traditional staff and more recently recruited "progressive" staff, the Forest Development Office (DGDF) was created, aimed at promoting "social production."[6] This policy would later become an important element in the evolution of the peasant forestry sector.

Aided by the patronage of Cuauhtémoc Cárdenas, the DGDF brought together a team of progressive technicians with a democratic vocation who were interested in walking a different path from that of the dominant group of conventional forestry engineers. The latter considered the only viable forestry production to be one based on the private sector, guided by the principle of productive efficiency. Yet several NGOs found an opportunity for negotiation within the DGDF,[7] and some NGOs even collaborated temporarily with the DGDF. From 1977 until the mid-1980s, under the flag of "social production," the DGDF undertook an extensive program of awareness-raising, training, and organizing. This led to the creation of several Forest Raw Material Production Units (UPMPs).[8]

The UPMP concept postulated that the devastation of forest resources could be stopped and reverted only through the active participation of the actual owners in the logging, management, and transformation of their resource, generating income and employment. Beginning in 1982, under the initiative of the DGDF and the UPMPs, several communities in Quintana Roo were finally able to obtain logging permits directly and to market their timber without the intervention of the contractors.

During the late 1980s, the effort of peasant organizations and government technicians was translated into a series of laws favorable to the peasant enterprises. According to the 1986 Forest Law, only the actual owners of a forest can be given the concession for its exploitation: forestry communities could officially become loggers and potentially evict external extractors.

Over that time, in southeastern Mexico, two institutions with an approach similar to that of the DGDF markedly influenced this legal and practical change: the Pilot Forestry Plan (PPF),[9] created in 1982 through a bilateral technical aid agreement between Germany and Mexico,[10] and the Mexican-German Agreement (AMA), instituted in 1986. Being bilateral programs, both had managed to gain sufficient political space for their activities and for their staff to practice an approach distinct from the conventional forestry research, planning, and development programs hitherto applied.

The PPF itself had only limited experience in Mexico before moving into the Peninsula. In the late 1980s, the PPF concept of autonomous community forestry was first applied in the state of Chiapas. However, it faced strong opposition from two successive state governors, because it questioned the corrupt organizations of the government party (PRI) that handled the processing and marketing of secondary products such as railway sleepers. When the PPF had to move out of Chiapas, it found in the state of Quintana Roo an extraordinary room for maneuver, by virtue of several circumstances:

- The international criticism of World Bank–financed deforestation, which was linked to the promotion of extensive cattle ranching in the tropics.

- The end of concessions to private logging companies.

- The mobilization of several peasant organizations.

- The political will of the DGDF staff.

- The interest of the new governor of the state of Quintana Roo in the project.

Trying to stop deforestation, the PPF's efforts were aimed at designing methods to support the social and economic role of community sustainable forestry in twelve *ejidos* in the south-central part of the state of Quintana Roo.

The Role of Innovative Rural Technicians

The favorable opportunities created during the 1980s in the governmental forestry sector for peasants' participation led to several technicians, from both within and outside governmental institutions, providing administrative and technical support to the forest producers, using a self-management approach. The long duration of some of these experiences permitted a close collaboration between technicians and peasants, imbued with confidence and solidarity.

Usually, the role of these external agents was not restricted to promotion, technical assistance, or the production of inventories and management plans. It also included maintaining the flow of information to and from the *ejidos*, managing the projects, looking for commercial contacts, promoting an organizational linkage at community and regional levels, and other forms of support not always related to productive aspects. In some cases, these technicians became involved in the programs as individuals; in other cases, as members of NGOs whose objectives matched those of the governmental institutions.

The NGOs, despite the great diversity of their approaches, displayed some common elements. First, they were marked by the ideology of the social movements of the late 1960s and the 1970s. With socialist, humanist, or Christian ideals, the emphasis was always placed on solidarity with the people and grassroots participation. Second, regarding the rural environment, the mystification of peasant communities led to a number of failures, since it prevented the search for and implementation of realistic alternatives and made smooth relationships with state structures impossible. Trying to build "from below," eyes were kept on the micro level. There was neither an interest in nor an ability to dialogue with the state, political parties, or private enterprises. Everything that sounded like political control or marketing was treated with suspicion.

Those technicians who penetrated the governmental forest development institutions in order to support the producers' organizations had a different kind of experience. This was the case among those who collaborated with the PPF in Quintana Roo. As mentioned earlier, this technical team was created with governmental support but had an unusual autonomy from the traditional bureaucratic apparatus and local logging companies.

The principles of the PPF, then, were the same as those of the DGDF: the devastation of the rain forest could be stopped only by means of the active participation of the owners in the logging, managing, and transforming of their resource. This hypothesis opposes the traditional perception of wood as natural capital to support an enclave industrial economy and of resource owners as hired hands within the process of exploitation. Moreover, the PPF developed a holistic concept of sustainable development. According to this concept:

- A development project cannot be predetermined.

- Natural, social, and economic components of development do not necessarily evolve at the same pace.

- Even though technological aspects are very important, forestry activities depend on market conditions for timber and nontimber forest products; on government policies at federal, state, and local levels; on the interest of the private sector; and on the cultural history of peasants.

- Social processes have to be granted a privilege within the action, since they are the only means to stop the devastation of the forest.

The young technical team of the PPF emerged in part from a critical generation of the agricultural university of Chapingo near Mexico City. It was led by a forestry engineer with expertise in the management of tropical forests, who had managed to overcome the opposition of the old government technicians. The PPF team was joined by anthropologists, economists, and a German technician.

A Technical Forestry Office (DTF) was created within every regional

producers' organization. When conceived by the PPF, the DTFs were supposed to receive their own financing from the organizations themselves, but in
most cases, this proved to be unfeasible. Several technicians were assigned to
the DTFs by the federal Forestry Department, which meant that they had to
accept an uncertain income and employment situation. They could stay there
only by virtue of their personal efforts to get additional financing from some
governmental institutions and the Mexico-Germany bilateral agreement.

Even though the DTFs were subject to the pressures of community members, regional interest groups, and government institutions, they made possible a forestry service in the hands of the peasants. Today, however, the
financing of the technical services is still an unsolved problem, with small-
scale forestry producers accounting for little more than 30 percent of the
DTFs' budget.

In short, the progressive (governmental or nongovernmental) technicians
fulfilled an important role in the genesis and consolidation of the peasant
forestry enterprises. Although they had to challenge the traditional forestry
service, they built up the ability of the small-scale producers to manage their
own resource. In this process, the technicians themselves went through a
learning process, since initially, many of them had limited technical knowledge and only a certain amount of social sensitivity.

Concepts and Work Strategies of the PPF:
The Experience of Quintana Roo and Southern Campeche

The unorthodox approach of the PPF—conceiving of issues well beyond the
sheer technical elements—received extensive criticism from the traditional
forestry bureaucracy and, later on, from the German agency that financed it
(GTZ).

It was indeed a difficult process. First, it was necessary to build consensus
among the peasants about the need to fight for their natural resource. This
was not too difficult, taking into account their experience with the private
logging companies that had caused nonselective deforestation. Then, some
incentive had to be found to get the people involved with enthusiasm in
communal forestry (Bunch 1986). This was certainly a difficult process,
requiring a counterbalance to the law of idle lands (*tierras ociosas*), which
effectively promoted deforestation; corrupt peasant organizations; pillaging
by logging companies; and convoluted bureaucratic structures that stimulated corruption.

Achievements and Difficulties

The instrument for counterbalancing these factors and encouraging peasants
toward sustainable forest use was, from the beginning, the valuation of the

forest. This led to one of the main achievements: the nearly threefold increase in cedar and mahogany prices, as a result of negotiation with the timber companies. In addition, one of these companies agreed to buy not only tropical fancy woods but also ordinary softwood and hardwood.

The next task was to convert the free-access model used by the private logging companies into a controlled-access model to be used by the new peasant social enterprise. An ecological and social land-use plan was designed using consensus-seeking methods, and the concept of permanent forest areas was introduced. This was followed by the adoption of an ordered forestry routine. For the first time, the peasants learned to measure the logs with the metric decimal system, a fairer method than that imposed by the companies (Doyle system), and to complete the required documentation. These apparently banal instruments are important for forest producers, since logging companies used them to cheat peasants and bribe their leaders. This training was followed by further technical adaptations, using only the knowledge inherited from the logging company. The territory was divided into twenty-five annual harvesting blocks, to be harvested consecutively during a rotation cycle of twenty-five years. After three cycles, or seventy-five years, mahogany trees reach an adequate size to be used again for commercial purposes.

This technical management approach had to be refined, however. The inventories—a technical instrument for management plans—for example, used to be carried out on the basis of the indigenous Mayan population's knowledge about soils and forests but this was gradually refined. This method of gradual approximations and mutual learning was a compromise between what was desirable from the ecological point of view and what was possible at any given stage of the learning process. Moreover, over time, it became noticeable how the technical studies were enriched with traditional knowledge, whereas the local population could evaluate with Western precision their own resources.

The process of negotiation with the timber companies strengthened the organization of the communities. With the funds saved in the National Ejido Promotion Fund, some logging machinery (now obsolete) was purchased, which led to further technical strengthening of the organization. Moreover, the decision-making structures, such as the general assembly, were reinforced, and administrative controls were established. Yet a controversy over profit targets began and continues today: some producers wanted immediate profit distribution, and others preferred reinvestment with the aim of generating more permanent work opportunities.

But political conditions continued to affect the work of the PPF. When the changeover of the state governor approached in 1988, the whole project was put at risk. A structure had to be found to ensure its continuity, to organize collective marketing, to negotiate standardized prices, to consolidate the DTFs, and to prevent the regional organization from being co-opted by the corporate peasant sectors. In this way, several NGOs were created in the legal form of civil society associations, *sociedades civiles*,[11] offering several advantages to the emerging social enterprise sector:

- They brought together *ejidos* and their interests around the concept of communal forestry, avoiding unrelated and individualistic action. They functioned as transmitters of information about relevant markets, projects, and funds.

- They hosted the forestry service, where sustainable silvicultural policies and practices are defined and where information is processed and disseminated to the *ejidos*.

- They offered space to the technical groups that promote projects working on nontimber forest products, agriculture, and so forth.

- They were a negotiating front for forestry policies. This is an extremely important role, since institutions such as the Regional Council represent a social power that adversaries cannot demolish or ignore easily.

- They represented a common front to negotiate market prices for forest products.

- They provided a framework to channel funds and subsidies from the federal government and NGOs to the *ejidos*.

- They provided *ejidos* with some services not covered by the government (for example, they own and hire plant machinery and mobile sawmills).

- They supported microindustries created by groups of peasants.

- They were a training ground for parliamentary and negotiation techniques and for leadership in forestry issues.

Spreading the Experience

This scheme of organization and technical approach was extended throughout much of Quintana Roo between 1989 and 1994 and was reproduced in the neighboring state of Campeche in 1990. Yet even though the forests of Quintana Roo and Campeche are similar, the size of the *ejidos*, the makeup of the timber resources, and the social history of the inhabitants (either settlers or Mayans who resisted the settlements) are very different. Therefore, the nature of the scheme had to vary when it was transferred from one place to another. Briefly, the three main adaptations of the original ideas, with successes as well as limitations, were as follows.

Southern Quintana Roo. Although this area has a significant volume of mahogany wood (the guiding species for logging), it is just a fraction of the total volume of hardwood and softwood available in this diverse forest.[12] Unfortunately, there is a market for only a few species, and since 1988, timber imports from Belize and Guatemala have forced market prices down. In this context, the sale of mahogany alone cannot provide sufficient income for a growing population, and new approaches are required.

In the beginning, the strategy of the PPF was to establish permanent forest areas and to undertake inventories. There followed the promotion of nontimber forest products, such as chicle (a resin used as a base for chewing gum), honey, and fauna (requiring a strong discipline in the management of the commons), and the sale of noncommercial hardwood and softwood. Today, there are good prospects for the opening up of green-seal (certified) timber markets.

Many *ejidos* already have sawmills, but the processing pattern that still prevails is one in which the resource is adapted to the requirements of the industry (especially that of plywood) and not the other way around. Under this pattern, an optimal forestry design is not possible, and a large proportion of the resource is wasted.

Several *ejido* enterprises have been set up, but their management structure is still inefficient. First, they are managed by the president of the *ejido*, a man who is usually a better politician than administrator and is too busy with other activities. Other positions in the social enterprises are assigned by the *ejido's* general assembly, but their criteria usually have more to do with power relations than with personal abilities. Second, most of the profits are distributed instead of being reinvested. Finally, participation of women is notoriously lacking.

Central Quintana Roo.   This area, with a large Mayan population, has suffered extensive nonselective deforestation of precious woods (much more so than in the southern part). Its main products so far have been railroad sleepers and chicle (Janka 1981).

When the PPF scheme was established here, two large *sociedades civiles* were constituted; one of them, the OEPF-ZM counts as members the majority of the 8,000 *ejido* forest farmers. As in the south, the first step taken was the definition of permanent forest areas, followed by the formulation of a forestry plan centered on the exploitation of sleepers. The latter, however, faced many problems, largely because the extraction of raw materials for, and final manufacturing of, sleepers has commonly been an individual activity, generating some income for the poorest people. In addition, social and economic structures of the market tend to obstruct the implementation of a disciplined management plan.

Taking into account the complex network of vested interests in the chicle market (especially that of the governmentally controlled Federación de Cooperativas), the PPF decided not to become involved with this product.

Peasant organizations have also had many disagreements in this respect with the regional government, but they are now succeeding in opening direct links into the market, bypassing intermediaries, and benefiting from tariff-free export conditions. Reforestation activities to avoid senseless deforestation in areas where chicle functions as the guiding species have not yet been well established, however. In fact, the *ejido* with the highest volume of mahogany within this region (X-Hazil) has left the organization.

Other successful strategies of the peasant organizations in central Quintana Roo have been the creation of a market for organic honey (with a price two times higher than that of standard honey),[13] the establishment of communal mobile sawmills (OEPF-ZM 1992), and the promotion of craft making among approximately 200 women who design clothes with rainforest motifs that are of interest to the tourists in Cancún.

Today, the main challenges faced by the peasant organizations of central Quintana Roo are how to promote more effectively an agroforestry system for plots in secondary forests (usually 20 hectares per family), how to modernize the technology employed in the important task of beekeeping, and how to contain the rapidly advancing breakup of permanent forest areas into smaller lots—a dangerous process in some of the *ejidos*.

Calakmul, State of Campeche.   The main peasant organization in this area is the Regional Council of X'pujil (CRASX), incorporating forty-four of the seventy *ejidos* in the buffer and inner zones of the Calakmul Biosphere Reserve (CRASX 1994). In 1991, with the changeover in regional governors, parties to the AMA and the Social Development Ministry (SEDESO) began lobbying the new governor, defending the need to link development with conservation activities. Since, at the time, these regional political personalities were supporting the influential presidential candidate Colosio, the political protection received until his assassination allowed them a period of relative political calm until the middle of 1995. After one of the council's advisers (and also a member of the Mexico-Germany Agreement) was appointed director of the reserve, the links between both the regional council and the reserve began to bear fruit. As a result of the upheaval over the changes in governorship and the weak political cover now available to the region, the success of the work in the Calakmul area depends intensely on the strength and independence of peasant organizations and the NGOs active in the region.

This particular area, both within and bordering an important biosphere reserve, is itself a heterogeneous area, with *ejido* surface areas ranging from 60,000 hectares (the old chicle *ejidos*) to 2,000 hectares (the new settlements). In addition, some of the *ejidos* have suffered, and continue to do so, from illegal logging by large companies. This means that the strategies of the peasant organization have had to vary according to the conditions of each *ejido*.

The selective deforestation of most *ejidos* has led to a twin-track strategy: first, the promotion of the market for noncommercial hardwood and softwood, and second, the production of high-quality products (such as furniture, crafts, and implements for beekeeping). The latter process is carried out within the currently small volume of timber available and represents an important processing activity to small *ejidos*, because of their control over the whole production chain. In addition, the regional council has promoted a multiple land-use strategy that includes agriculture, agroforestry, nontimber forest products, cattle raising, and home gardens (Boege and Murguía 1989).

Turning the present slash-and-burn agricultural system into a more sedentary land-use practice is one of the main objectives throughout the region. Taking into account the low quality of soil for the required agricultural productivity, this objective is being achieved by means of green manure or leguminous cover crops that maintain soil humidity and help increase fertility. At present, 500 of 2,000 peasants are using these techniques.

In 1,200 hectares of young secondary forests, agroforestry techniques are now applied that include precious wood species for long-term production as well as fruit trees—especially citrus fruit—and food crops for annual production. As in the other two regions, the main nontimber forest products are chicle, fauna, allspice, and honey. Honey production has had to overcome both the invasion of African bees and the corruption of the established marketing agencies. By now, however, the high price of honey has allowed its production to become one of the main means to imbue permanent forest areas with economic value—as feeding grounds for the bees.

The Flexible Approach of the PPF

In southern Quintana Roo, the PPF and the local *ejidos* introduced the concepts of peasant participation, land-use planning, permanent forest areas, wildlife management, and harvesting of some nontimber forest products such as honey and chicle. When applied to the Mayan central part of the state, the strategy included some management of the extraction for sleepers, the certification of honey, and craft making. Finally, in Calakmul, emphasis was placed on multiple-land-use and low-external-input agriculture. The common characteristic of these strategies is the diversification of local economic activity, coupled with an enhancement of economic and ecological efficiency.

In the final analysis, the noted high degree of collaboration between different actors over such controversial issues as the management of tropical forest resources has as much to do with particular individuals as with particular political opportunities. At the same time, the experiences of peasant organizations in Quintana Roo and Campeche have also had an impact on and have received feedback from collaborating networks of NGOs, national peasant organizations, and progressive sectors within governmental institutions.

Exchange and Systematization of Forestry Experiences

People involved in the processes of peasant forestry, either as producers or as advisers, have effectively entered a new field that as yet has few external reference points. The main source of knowledge for the actors is their own practice. Unfortunately, in this environment, there is little tradition of systematizing and sharing such experiences. Activism is so hectic that a reflection about one's own practice takes place only once in a while, and usually as a response to funders' requirements.

In the late 1980s, trying to fill this gap, a network of NGOs that had been involved in the support of peasant organizations for over a decade launched an innovative program aimed at the exchange of experiences and discussion about working methods, principles and practices of involvement in the rural arena, and forms of relations with peasant groups. This program, called PASOS ("steps," in Spanish), undertook a collective and critical analysis of the following issues:

■ The internal dynamics of peasant communities and organizations.

■ The performance of NGOs and working groups.

■ The relationship between technical teams and rural populations.

■ The relationship between funders and other stakeholders.

One of the program's activities was the organization of workshops for the analysis of forestry experiences from the perspective of several NGOs that had been working with peasant groups on forestry issues, providing technical and administrative training and giving advice on organizational processes. Some of these NGOs are the Environmental Studies Group (GEA), Rural Studies and Advice (ERA), and Alternative Services for Education and Development (SAED).

The three workshops organized between 1989 and 1992 were also attended by members of peasant organizations and academic institutions.[14] The following list of questions gives an idea of the issues that were dealt with:

■ When a contradiction arises within a peasant enterprise between the need to make decisions democratically and the need to operate efficiently, how is it unraveled?

Activities of CCMSS

Multisector meetings
Case studies
Participation in national and state technical consultative councils
Support to peasant organizations
Training
Participatory formulation of management plans
Cartography and calculations
Timber certification

- How can the conflict between a rationale that requires competition in the market and the peasant rationale be solved?

- What are the prospects of the peasant enterprises in the context of the North American Free Trade Agreement (NAFTA)?

- Is the function of NGOs to substitute for the state?

- How can the subordination of NGOs with respect to their funders be prevented? How can true cooperation be achieved?

- Are NGOs at the peasants' service, or do they maintain their own identity and objectives?

The discussions helped the participants (NGO members, academics, and technical staff of the producers' organizations) see their weaknesses and strengths, formulate common objectives, and visualize the prospects for common action.

At that time, an enhanced ability to dialogue with other parties (with government bodies, other institutions, national and international networks, and so forth) was observed. Apparently, this advance was due, among other factors, to the independent technicians' ability to make proposals, to the opening of opportunities for participation in some government institutions, and to the influence at the international level of the discourse and the policies proposed at the 1992 United Nations Conference on Environment and Development in Rio de Janeiro.

As a result of this exchange of experiences, the need to undertake joint action was observed. This led to the creation of the Mexican Civil Council for Sustainable Forestry (CCMSS).

CCMSS was constituted in May 1994 by the participants in the PASOS workshops, as well as by other NGOs and individuals interested in the promotion of sustainable peasant forestry.[15] CCMSS actions include the following:

- Several multisector meetings aimed at achieving a better understanding of the social, economic, and environmental implications of sustainable forestry have been organized. An increasing diversity of social sectors, including the private sector, has been included.

- CCMSS has participated in several committees for the design of forest policies: the National Forest Technical Consultative Council and the consultative councils of the states of Jalisco and Veracruz.[16]

- Several case studies have been carried out with participatory methodologies in order to evaluate the forest exploitation in several regions of Mexico, to identify key problem areas, and to devise ways to make forestry more sustainable (Alatorre, Merino, and Gerez 1995; CCMSS 1996). The first objective is to define with more clarity what sustainability means in the context of Mexico.[17]

- The methodology used for these studies is analogous to that of the participatory rural appraisal (PRA). It stems from an explicit demand made by a community or an organization and includes a series of field interviews and several discussions with groups of peasants. In a final feedback session, identified problems are prioritized, and a work schedule is drawn up by the village assembly.

- Several peasant organizations receive support (contacts and funds) to carry out training courses, forest studies, and management plans. Moreover, CCMSS promotes the training of its own members.

- CCMSS is promoting a process of certification and the opening of a market for certified timber. In December 1994, the council and SmartWood (a U.S. association) certified the timber produced by UZACHI, an organization of the state of Oaxaca.

- CCMSS has participated in the conception of the Forest Stewardship Council and has established links with national and international networks, which may increase the possibilities of influencing the design of public policies to stimulate sustainable forestry (by means of market mechanisms, economic or fiscal incentives, and so forth).

One of the tasks of CCMSS is the dissemination of experiences and opinions. It publishes a bulletin and maintains an electronic conference. Recently, it opened a web page.[18] Moreover, by means of PASOS's membership in the international Dialogues for the Progress of Humanity (DPH) network, there is a link with over 100 organizations and documentation centers around the world that make up this network, originally promoted by the French Foundation for the Progress of Humanity. The DPH node in Mexico is GEA, an NGO that since the late 1970s has been working in communal forestry. Since 1990, GEA has also provided training to groups and individuals that want to use the DPH system for a systematization and sharing of experiences.

PASOS also manages a database. Both DPH and PASOS databases work by the principle of barter: they are available to every organization that agrees to share its own experience. The exchange is usually carried out by sending diskettes through the mail. In addition, the data cards that constitute the bank of experiences have been disseminated by other means, such as ad hoc publications produced as inputs for meetings and seminars and papers in the PASOS newsletter. PASOS and the members of CCMSS are in touch with other Latin American organizations by means of electronic networks, such as the Rural Information Network in Mexico and the Agriculture and Democracy Inter-American Network (RIAD).[19] These electronic networks host a rich exchange of experiences that are usually also recorded as DPH cards.

One of the main challenges facing CCMSS today is how to get closer to the producers' organizations. If NGOs are not used to systematizing their own experiences, in the case of peasant organizations, this lack is even more apparent. A new culture of producing and circulating knowledge may well help the latter face the new challenges ahead.

Peasant Forestry at the Time of NAFTA

In the 1990s, forestry production by the peasant sector (the social forestry enterprise) in Mexico faces a series of opportunities and risks. Even though in most cases it will not meet all the expectations of communities and organizations in terms of income and sustainable management of resources, its ability and power to dialogue and make proposals are much more relevant than some time ago, when national and foreign logging companies had an absolute predominance.

Even before NAFTA, the Mexican forestry sector faced serious marketing problems, on account of:

■ Mexico's entrance into the General Agreement on Tariffs and Trade (GATT) in 1986, which opened the borders to the import of very cheap timber.

■ A high inflation rate that shrunk the internal market.

■ The recession, which choked the building and furniture industries.

■ The reduction in public expenditure, which led to a reduction in public works (which demand large volumes of timber).

Mexico's entry into GATT took place at a time of national economic crisis, with high inflation and high interest rates. The attractive returns of the stock market discouraged productive investments, and the opening up of the borders in this context led to a number of insolvencies of national enterprises. Moreover, the economic measures implemented since 1988 to control inflation, keep a fixed exchange rate, and increase financial rates of return led to an overvaluation of the national currency and thus to a further negative price balance between national and imported products. Between 1988 and 1994, this led to a much faster rise in imports than in exports. NAFTA, which came into effect in 1994, simply meant the continuity of a policy that had been outlined since 1985 (Oswald 1992; Szekely 1994).

This picture changed with the devaluation of the Mexican peso in December 1994. However, the ensuing contraction of the internal market paralyzed the building and furniture industries, as well as the purchasing power of the national consumers. This had a very strong effect on Mexican forestry: the demand for timber products diminished, and market prices had to be reduced concomitantly. On account of the much lower production costs of timber in the United States and Canada,[20] the cost of Mexican coniferous timber is 1.66 times higher than that of equivalent timber in the United States (Chapela 1992, 350; Merino 1992). Moreover, since the late 1980s, no credit has been given by development banks to increase the efficiency of the Mexican forestry and timber industries (Cortéz 1992).

Social forestry enterprises, however, have shown a greater ability to withstand the crisis than their counterparts in the private commercial sector. In the state of Oaxaca, for instance, a number of private enterprises closed when their rates of return shrank. Capital can freely move to other branches of industry or, given the high interest rates, be used as a financial investment. However, the rationale of the social enterprise sector is not the same: as long as employment is maintained, the returns can be sacrificed; its resilience is greater than that of private enterprises.

Although not all of them can prevail through such hard times, several peasant enterprises in different places are thriving with regard to forest management, industrial efficiency, and administration. Due to the experience they have accumulated, there are now an increasing number of technicians from the communities themselves who are fulfilling the tasks previously accomplished by external advisers.

Another factor in the political strengthening of communal enterprises is the links these businesses have established with one another. These relations have led to the creation of an identity as a different form of producer organization, to a number of peasant producer meetings, and to the constitution of networks such as the Mexican Network of Forest Peasant Organisations (MOCAF), which since 1990 has been bringing together *ejido* unions of several states.

With respect to changes in the legal framework, there has been both profit and loss. On the one hand, the new forestry law that came into effect in 1992, in conjunction with reforms to the agrarian law, promoted the neo-liberal model by opening up the exploitation of the forests to private investment and enabling the concentration of forestland in the hands of private societies. On the other hand, the same forestry law diminished the requirements of documentation and enabled private consultancy firms—instead of governmental technical teams—to take on the technical responsibility for the extractive activities. Even forest *ejidos* and communities have been able to take advantage of these changes.

Broadly, the impact of communal forestry can be observed at different levels. In the environmental domain, the case of southeastern Mexico is significant. In the social, economic, and political domains, the following aspects can be highlighted:

- At the local level, there is an improvement in the quality of life of the inhabitants, including more basic services (drinking water, electrification, roads, and communication); greater employment opportunities (reduced need to migrate to the cities or abroad); more agricultural and cattle production (more resources to invest); and better management abilities (better ability to dialogue with other political and economic actors). It can thus be said that communal forestry has substantially contributed to the overall strengthening of the communities.[21]

- At the regional level, the organizational, economic, and political strength of several unions of forest *ejidos* or communities has been enhanced. Their ability to control production, processing, and marketing and to provide forestry technical services has gradually been built up. Finally, their economic strength enhances their political power.

- It is at the national level where the weaknesses of communal forestry are underlined: it has not been possible to coordinate isolated efforts (for example, by means of a federation) in order to increase the lobbying power over the policymaking process in the forestry sector.

Challenges of Peasant Sustainable Development in the Southeast

The concept of sustainable forestry developed in the states of Quintana Roo and Campeche on an area of 500,000 hectares of peasant-established permanent forest areas. It was shaped by the Mexico-Germany Agreement and then facilitated by a series of favorable opportunities. A range of peasant organizations were created, and they are now strong enough to survive without external support, or at least without support of the respective state governments. Even though these organizations do not cover all the forest area (several *ejidos* are still disorganized, or their resources are too scarce), there are already enough peasant experiences to be able to devise regional policies for sustainable development of the region's rain forest.

The role of the mediators has been essential in this process. The Mexican political system, under which new administrations are set up along with presidential elections every six years, implies a discontinuity in most of the programs. In this context, the long-term presence of these mediators-advisers has meant a solid "moral authority." Mediators fulfil—and have to keep fulfilling—the role of broker or interface between peasant producers and the outside in several domains: political, organizational, technical, economic, and commercial.

Both producer organizations and their mediators face a number of urgent challenges—namely, to generalize those experiences that have been successful; to adapt the trialed model to specific situations; to provide technical, social, and economic support to some weaker processes; to generate a competitive entrepreneurial approach; and to promote the creation of more employment opportunities by means of forest production and the processing of timber.

Challenges and Strategies

In the light of the preceding discussion, the following strategies appear to be open to both mediators and social forestry enterprise:

1. In areas with adequate potential for timber production, the harvesters need to turn into foresters. This involves managing forest patches, ensuring the regeneration of cedar and mahogany, developing appropriate and low-cost technologies for extraction, and carrying out inventories and studies geared toward a multiresource approach. In the process of extraction, mobile sawmills have a promising future, since they enable tips and branches to be sawed and avoid the need for vehicles to transport raw logs to the sawmill.

2. Taking into account that the DTFs are one of the most important instruments for technical, organizational, and administrative development, a regional approach requires training new recruits and teams. Also important is the development of mechanisms for financing the sharing of skills and experiences, in addition to technical and political forums and debates about specific problems. A regional organism coordinating all the technical offices would be desirable, incorporating both empirical and specialized technicians. Moreover, the DTFs need to achieve some degree of independence with respect to the peasant organizations without breaking well-established linkages. Finally, the financing of technical services has to be solved, since current profits are not high enough to ensure their provision.

3. The pattern of communal organization needs to become one of a social enterprise, since the former, despite all its virtues, has several shortcomings. Even though the managing paradigm of social enterprises is not yet clearcut, there has already been some success with enterprises created in large *ejidos* whose members are shareholders but not administrators. They have realized that if all the financial partners have to take part in the decision making, the social enterprise becomes too clumsy—too many captains and

not enough sailors. Technical and administrative skills have to be developed within the *ejidos* and separated from the political decisions and positions; also important is the creation of accounting systems and managerial forms appropriate to the conditions of peasant enterprises. Last, the growth of the *ejido* population should not lead to a reduction in profits. This is why it is essential that new positions (especially for women) be created, which requires that profit redistribution be limited. This demands a high level of confidence in the enterprise.[22]

4. One of the main challenges is the creation of a joint institution that coordinates the marketing of timber, handles quality control, manages the stocks of quality timber (dried and cut according to customers' desires), and brings together the small volumes produced by the *ejidos* to offer significant volumes of timber to the market.

   Timber-processing industries should also be developed, in order to increase value-added on local production. This includes sawing, precision sawing, drying, and manufacturing end products in workshops or factories. This industry has to relate the resource to its owners: for instance, it has to collect small-sized timber (tips, branches, and debris). Moreover, it has to tailor the local market, especially that of national and international tourism in the nearby Cancún-Tulum corridor. This means designing craft objects, producing them in a set of coordinated microenterprises, and selling them with a "green stamp," certifying the sustainable use of forests and the well-being of the population.

5. Regional economic planning should be fostered, including archaeological and ecological tourism, together with adequate management of the biosphere reserves as economic opportunities. The mechanisms for the international transfer of resources to repay some environmental services of the forest, such as carbon sink capacity and biodiversity conservation, are already being established. These resources could be used to finance some uneconomic activities that community forestry cannot sustain, such as roads, research, inventories, clearing of gaps for the regeneration of cedar and mahogany, and the DTFs.

Mediators and NGOs at the End of the Millennium

The context of NGOs is very different in the 1990s than in previous decades. Both NGOs and peasant organizations have changed. There is less ideology and more pragmatism; advisers face the challenge of building, along with the peasant communities and organizations, a politically, economically, and environmentally feasible project of community forestry. This path cannot be followed simply on the basis of goodwill and voluntarism, but must be the result of negotiations and transactions between different stakeholders.

Today, NGOs are taking on a new role that, without breaking the direct or indirect relationship with producers, incorporates other dimensions, such as the relationship with the state and different markets and the creation of links with governmental bodies and academic and international institutions. On account of their ability to act as "cultural translators," those NGOs that advise on forestry issues have already responded to the need for bridges between the peasant world, civil organizations, academic or governmental institutions, and the international aid community. They have also contributed to bringing together the spheres of production and environment, of praxis and theory, and of the local and the global. The existence of entities such as the forest technical consultative councils for the design of public policies for sustainable forestry management, and the participation of CCMSS within them, constitutes an opportunity to reinforce the evolving process of professionalization and strengthening of community enterprise that is sought by many.

Notes

This chapter was translated from Mexican Spanish by Claudio Alatorre.

1. There are three kinds of landownership schemes in Mexico: *ejido*, agrarian community, and private property. Every *ejido* has a given number of members that have usufruct rights but do not hold property titles. Before the 1992 reforms to the agrarian law, *ejido* plots could not be sold, rented, or seized, in the spirit of the agrarian revolutionary struggles of the early twentieth century. In reality, however, the *ejido* agricultural land is usually divided up into individual plots, whereas forest or pasturelands are usually collective property. The agrarian communities (*bienes comunales*) are based on a similar regime but apply to indigenous communities with property titles dating from the Spanish colonial times.
2. This means the exploitation of precious woods without any management plan.
3. Snook (1986, 27–31) describes the fight of some communities in the state of Oaxaca against the regranting of the concession given to the Tuxtepec paper factory.
4. Examples of regional forestry organizations are the Unión de Ejidos Emiliano Zapata, in Durango; the Unión de Ejidos del Noroeste and the Unión de Ejidos Francisco Villa, in Chihuahua; the Unión de Permisionarios de la Meseta Purépecha, in Michoacán; the Unidad de Producción Adalberto Tejeda, in Veracruz; and several foresters' *sociedades civiles* in the southeast and several organizations in the state of Oaxaca, such as the Unión Zapoteca-Chinanteca, the Unión de Comunidades Forestales Ixtlán-Etla, and the Unión de Comunidades y Ejidos Forestales de Oaxaca (Lara 1992).
5. See Alvarez I. et al. 1992, especially the article on the Sociedad Civil Forestal de la Zona Maya de Quintana Roo (pp. 93–114). Lara (1992) describes the experience of Huayacocotla.
6. "Social production" is a Mexicanism meaning civil society, or local producers controlling production and possibly manufacturing processes, as opposed to external corporate interests.
7. The DGDF was headed by a renowned engineer, Jorge Castaños.
8. Depending on the peasants' previous organizational experiences, local leadership,

and the approach followed by the government in their promotion, some UPMPs actually became peasant enterprises, while others maintained their nature as government programs.

9. For a full description of the PPF's history, see Lanz, Argüelles, and Montalvo 1995.

10. Carried out by the German technical assistance governmental agency GTZ.

11. The following *sociedades civiles* were formed: the Sociedad de Productores Forestales Ejidales de Quintana Roo; the Organización de Productores Forestales Ejidales de la Zona Maya, S.C.; the Sociedad de Pueblos Indígenas Forestales de Quintana Roo Tumben Cuxtal; the Organización de Ejidos Forestales de Quintana Roo-Chaktemal; and the Consejo Regional de X'pujil, in the Biosphere Reserve of Calakmul, Campeche.

12. The approximate volume of exploitable hardwood per hectare is 80 cubic meters in Calakmul, 100 in Quintana Roo, up to 200 in Chiapas, and, as a reference, 250 in some forests of Southeast Asia. Out of the 80 cubic meters in Calakmul, the eight commercial species amount to only 10 cubic meters, and just a few *ejidos* exceed 1 cubic meter per hectare of cedar and mahogany. In comparison, a temperate forest has more than 350 cubic meters per hectare of exploitable wood (Deocundo Acopa, personal communication).

13. In 1993, ordinary honey was sold at US$0.70 per kilogram in the United States (organic honey, US$1.40). Then the United States imposed a tariff on Chinese honey, causing the price for ordinary honey to rise to $1.40.

14. The proceedings of these workshops (held in July 1989, November 1990, and November 1991) were published by PASOS (see PASOS 1990, 1991).

15. Besides ERA, GEA, and SAED, organizations that are part of PASOS, the following NGOs participate in CCMSS: the Centro de Apoyo al Movimiento Popular Oaxaqueño (CAMPO), the Proyecto Sierra de Santa Marta (Veracruz), the Grupo Interdisciplinario de Tecnología Rural Apropiada (Michoacán), and Educación Popular y Capacitación (Querétaro). Finally, there are about ten individuals who collaborate either independently or from within academic or governmental institutions.

16. The National Forest Technical Consultative Council is made up of representatives of governmental institutions related to the forest sector, national peasant organizations, national private-sector associations, and NGOs. Its tasks are consultation, evaluation, discussion, and proposition of forest policies. The consultative councils at the state level have analogous composition and tasks.

17. Taking into account that the Mexican forests are fairly densely populated, CCMSS is proposing the following sustainability criteria: compliance with all legal requirements, peasants' assurance of their land titles, sociocultural valorization of forests by their inhabitants, social benefits (such as employment) derived from forest exploitation, existence of communal mechanisms for consensus building and conflict resolution, economic feasibility of timber extraction, sustainability of the extracted volume, diversification of species and products being used, control of the environmental impacts of logging and extraction processes, and existence of a management plan governed by forestry and environmental principles, and the possibility to adapt this plan to different circumstances (which implies a monitoring process).

18. Access this page at http://www.laneta.apc.org/ccmss.

19. The RIAD network brings together individuals from NGOs and social organizations from Latin America and the United States. Created in 1992 with the objec-

tive of producing knowledge useful for the formulation of strategies and action by peasant organizations and NGOs working in rural areas, its central areas of interest are policies conducive to sustainable agriculture, peasant and indigenous organizations and local authorities, and regional integration processes (NAFTA, Mercosur, and so forth). The RIAD network participated in the organization of the International Forestry Meeting (Pátzcuaro, Michoacán, April 1993), which led to the constitution of the Interamerican Forestry Network.

20. Mexico has an inefficient industrial structure and much higher transportation costs due to the difficult topography of forest areas, which increases road-building and maintenance costs. The United States and Canada enjoy much better conditions, and they jointly produce 20 percent of raw timber in the world.

21. All the communal forestry enterprises we have heard about use some of their profits for road improvements, construction of schools, electrification, and other basic services that, in theory, should be provided by the state. In the specific case of Quintana Roo, the enterprise provides scholarships for higher education, as well as insurance in case of illness or death. The more successful *ejidos* (Nohbec, Tres Garantías, and Petcacab) have created retirement funds for old people.

22. In many *ejidos* and communities of the states of Oaxaca and Veracruz, all profits are used for community works and for the capitalization of communal enterprises, without any distribution of profits to community shareholders.

References

Alatorre, Gerardo, L. Merino, and P. Gerez. 1995. En la búsqueda de un manejo sostenible de los bosques: El Ejido Ingenio del Rosario, Xico. In *Alternativas al manejo de laderas en Veracruz*. Mexico City: SEMARNAP Fundación Friedrich Ebert.

Alvarez I., Pedro, G. Alatorre, et al., eds. 1992. Las organizaciones campesinas e indígenas ante la problemática ambiental del desarrollo. Universidad Autónoma de Chapingo, Mexico.

Bartra, A., M. Fernández, J. Fox, et al. 1991. *Los Nuevos Sujetos del Desarrollo Rural: Cuadernos Desarrollo de Base 2*. Mexico City: ADN Editores.

Boege, Eckart. 1995. *The Calakmul Biosphere Reserve*. Working Paper no. 13. Paris: UNESCO (South-South Cooperation Program).

Boege, Eckart, and R. Murguía. 1989. Diagnóstico de las actividades humanas que se realizan en la Reserva de la Biosfera de Calakmul, Estado de Campeche. Pronatura Península de Yucatán, Mérida, Mexico.

Bunch, Roland. 1986. *Dos mazorcas de maíz: una guía para el mejoramiento agrícola orientado hacia la gente*. Oklahoma City: World Neighbors.

CCMSS. 1996. *La forestería comunitaria en México y sus perspectivas de sustentabilidad: Nueve estudios de caso*. Mexico City: UNAM-SEMARNAP-WRI.

Chapela, G. 1991. De bosques y campesinos: problemática forestal y desarrollo organizativo en torno a diez encuentros de comunidades forestales. In *Los Nuevos Sujetos del Desarrollo Rural: Cuadernos Desarrollo de Base 2*, edited by A. Bartra et al. Mexico City: ADN Editores.

Chapela, Gonzalo. 1992. Hacia una plataforma para la competencia comercial en el subsector forestal. In *La disputa por los mercados. TLC y sector agropecuario*, edited by Alejandro Encinas, J. de la Fuente, and H. MacKinlay. Mexico: LV Legislatura / Diana.

Cortéz, Carlos. 1992. El sector forestal mexicano ante el TLC. *El Cotidiano* 48: 79–85.

CRASX. 1994. Bosque modelo para Calakmul. Ecología productiva. Propuesta, Consejo Regional Agrosilvo-pastoril y de Servicios X'pujil, S.C., Zoh-laguna, Campeche, Mexico.

De Ita Rubio, Ana. 1994. Notas para el análisis de la transición de las organizaciones campesinas ante un nuevo patrón de desarrollo agrícola mexicano. In *Economía: Teoría y práctica*. Mexico City: UAM.

González Pacheco, Cuauhtémoc. 1992. Los bosques y selvas de México, sus habitantes y las empresas sociales. In *El sector agropecuario mexicano frente al Tratado de Libre Comercio*, edited by Cuauhtémoc González Pacheco. Mexico City: UNAM, UACh, Juan Pablos.

Janka, Helmut. 1981. La alternativa forestal comunal: ¿una alternativa para el trópico húmedo? In *Acuerdo México-Alemania: Alternativas para el uso del suelo en áreas forestales del trópico húmedo*. Mexico City: SARH-INIF.

Lanz, Miguel, A. Argüelles, and F. Montalvo. 1995. The Society of Ejido Forestry Producers of Quintana Roo. In *Case Studies of Community-Based Forestry Enterprises in the Americas*, edited by Institute for Environmental Studies and Land Tenure Center. Madison: University of Wisconsin.

Lara, Yolanda. 1992. Posibles impactos de las reformas al Artículo 27 sobre los recursos forestales de México. *El Cotidiano* 48: 13–20.

Merino, Leticia. 1992. Contrastes en el Sector forestal: Canadá, Estados Unidos y México. *El Cotidiano* 48: 67–73.

OEPF-ZM. 1992. La experiencia de la Sociedad Civil Forestal de la Zona Maya de Quintana Roo. In *Las organizaciones campesinas e indígenas ante la problemática ambiental del desarrollo*, edited by Pedro Alvarez I., G. Alatorre, et al. Mexico City: Grupo de Estudios Ambientales–PASOS.

Oswald S., Ursula. 1992. El Campesinado ante el Tratado de Libre Comercio. *Cuadernos Agrarios* 2 (4): 42–59.

PASOS. 1990. *Memoria del Segundo Taller de Análisis de Experiencias Forestales*. Mexico City: PASOS.

PASOS. 1991. *La empresa social forestal. Memoria del Tercer Taller de Análisis de Experiencias Forestales*. Mexico City: PASOS.

Snook, L. C. 1986. *Community Forestry in Mexico's Natural Forests: The Case of San Pablo Macuiltianguis, Oaxaca*. General Technical Report SE-46. Washington, D.C.: USDA Forest Service.

Szekely, Miguel. 1994. Sobrevivencia de las Empresas Forestales Comunitarias en Sistemas de Mercado Abierto. Ph.D. diss., University of Wisconsin.

8

Organizing Across Borders

The Rise of a Transnational Peasant Movement in Central America

Marc Edelman

Since the late 1980s, peasant organizations throughout Central America have taken unprecedented steps to coordinate political and economic strategy. Farmers from the five republics that constitute the *patria grande* of Spanish Central America (Guatemala, El Salvador, Honduras, Nicaragua, and Costa Rica), as well as from Panama and Belize, have founded regional organizations that meet regularly to compare experiences with free-market policies, share new technologies, develop sources of finance, and create channels for marketing their products abroad. They have also mounted lobbying campaigns at summit meetings of Central American presidents and agriculture ministers, as well as at the World Bank and International Monetary Fund. In a striking sign of the organizations' rising political profile, the Central American presidents invited the main regional peasant association to submit position papers and to participate in their December 1992 agricultural summit, as well as to attend numerous subsequent presidential and ministerial meetings.

This chapter has three basic objectives:

1. To analyze the formation of the principal regional peasant organization, the Association of Central American Peasant Organisations for Cooperation and Development (ASOCODE), which has national member coalitions in all seven countries of the isthmus.[1]

2. To consider briefly the implications of this Central American case for social scientific theories of collective action and social movements.

3. To examine the extent to which the Central America–wide *campesino* association has transcended the traditional sources of weakness and division that afflict peasant organizations in this region and elsewhere.

In Central America in the 1980s, peasants were often participants in armed opposition movements (El Salvador, Guatemala, Nicaragua) or were locked in bitter struggles with governments over economic structural adjustment programs (Panama, Honduras, Costa Rica). But as the last decade's civil wars have ended or ebbed and as free-market policies have started to appear inexorable, small peasants' organizations have devoted increasing attention to carving out what they call the "economic space to appropriate the riches our production generates" (ASOCODE 1993c, 18). Lessened civil conflict contributed to grassroots peasant unity, as did changing notions of campesino identity, shifting geopolitics, and the availability of European "solidarity cooperation" funds.

Most importantly, smallholding peasants in different countries increasingly recognized that they faced similar, linked problems. These include:

■ Rapid steps by national governments and powerful entrepreneurial groups toward regional political and economic integration. This created new loci of decision making above the national states and threatened to leave grassroots sectors, particularly small agricultural producers, behind.[2]

■ The implementation of economic structural adjustment programs (SAPs), which slashed social services (including agricultural extension) and production credit, reduced farm price supports and other subsidies (such as for loans and inputs), reversed hard-won agrarian reforms, and facilitated the penetration of transnational capital in the agricultural sector (Fallas 1993; FONDAD 1993; Pino and Thorpe 1992; Stahler-Sholk 1990; Thorpe et al. 1995).

■ The reduction of extraregional tariffs, which required grain producers to compete with foreign farmers.

■ The liberalization of grain trade within Central America, which is already exacerbating sectoral and regional inequalities (Fallas 1993, 87–99; Solórzano 1994).

■ The rise in U.S. food aid, which glutted cereal markets and led consumers to substitute imported wheat for domestically grown maize (Garst and Barry 1990).

■ The collapse of coffee prices following the termination of the International Coffee Agreement.[3]

■ A severe environmental crisis, with growing agrochemical contamination of soil and water and a vicious circle of deforestation, diminished rainfall, erosion, and declining fertility.

■ The proliferation of new nongovernmental organizations (NGOs), often supported by "social compensation" funds from bilateral aid and multilateral lending agencies, which peasant organizations often view as interlopers and competitors in the field of rural development (CCOD 1990; Edelman 1991; Kruijt 1992).

■ The long-standing lack of access to transport, storage and processing facilities, and market information, which heightens peasants' vulnerability to and dependence on intermediaries and large-scale agroindustries and thereby lowers their incomes.

The phenomenon of internationalization also has roots in the efforts of a young generation of movement leaders who claim to embody and hope to propagate a new collective identity for the region's peasants. These activists—products of some two decades of regional upheaval, wars, and crisis—constitute a kind of peasant-intellectual (Feierman 1990) that has received relatively little attention from social scientists. Like other peasants, they have had to adapt to major technological changes in agriculture (first green revolution input packages and more recently high-risk nontraditional export crops), as well as to the complex demands of the financial, marketing, extension, cooperative sector and land tenure institutions, with which smallholding agriculturists increasingly have had to interact. Urban and rural cultures have also converged—and not just as a result of electronic communications media reaching into remote zones or migration from the countryside to the cities. In much of the region, a significant proportion of the economically active population involved in agriculture now resides in urban areas, and a growing portion of the economically active rural population is engaged in nonagricultural activities (Ortega 1992).

In some countries, particularly Costa Rica, growing access to higher education significantly expanded the cultural-intellectual horizons of a generation of young people, including many from rural, low-income families.[4] Elsewhere, peasant activists received courses from the cooperative movement, church and government institutions, political parties, guerrilla movements, NGOs, and campesino organizations. Even when they have relatively little formal education, many are well traveled, computer literate, and conversant in macroeconomic policy, national and international politics, and the latest developments in tropical agronomy and forestry.

Together with a small but highly committed core of pro-peasant intellectuals and technicians who work with their organizations, they aim to reshape the self-image of the rural poor, as well as notions about campesinos held by members of the larger society.

In particular, today's peasant leaders intend to replace the peasant's image as an atavistic rustic with that of a politically savvy, dignified, and efficient small producer.[5] Adept at appropriating and refashioning the dominant discourses about democracy and civil society, they claim to articulate an alternative—and more just—model of economic development and see themselves as active participants in the discussions about Central American regional integration. They have increasingly forged alliances with nonpeasant groups.[6]

Early International Contacts

In Central America, a region of small states and permeable frontiers, migration and participation in social movements abroad are not new for the rural

---

Training for Leadership

"I've had any number of courses," Amanda Villatoro remarked matter-of-factly. Born in 1961, she finished ninth grade in eastern El Salvador and went on to become a prominent leader of the Salvadoran Communal Union (UCS). "Statistics, microeconomics, macroeconomics, political theory . . . I have a long curriculum vita. The UCS helped train me, with very high-level professors, and although I never went to the university nor even finished high school, I believe the knowledge I've acquired is equivalent to the fourth year of [a university] economics [major]. These are the tools we need to interpret the numbers the governments and the business groups present us."

Jorge Amador, a leader of the National Rural Workers Central (CNTC) and the Coordinating Council of Peasant Organizations of Honduras (COCOCH), completed only one and a half years of high school, but he has had extensive specialized education: "I've participated in many training programs in Honduras, programs of the CNTC, such as a three-and-a-half-month program called technician in agrarian development, as well as other themes, sociology, a bit of philosophy, planning, agrarian law. I've also trained abroad, in Panama for three months. I've been in a great number of training programs and traveled to training events in Mexico, Nicaragua, Colombia, and England . . . . I've always taken the initiative to read a little, as much as possible, and I have a little library in my house."

Sources: Interviews in San Salvador, 21 July 1994, and in Tegucigalpa, 29 July 1994.

---

poor (Acuña 1993). Hundreds, perhaps thousands, of Nicaraguans, for example, participated in the 1934 strike against the United Fruit Company in Costa Rica (Bourgois 1989, 203). In Honduras, twenty years later, hundreds of Salvadoran banana workers joined their local counterparts in a massive walkout against the United Fruit–owned Tela Railroad Company (González 1978). Transnationalism—that circulation across borders of people, technology, money, images, and ideas, which has lately so fascinated anthropologists (Appadurai 1990; Kearney 1991)—has been well known to Central Americans for decades, if not centuries.

The 1970s and 1980s, nonetheless, saw an intensification and marked qualitative shifts in these transnational flows. In 1978–79, as the Sandinistas' campaign against the Somoza dictatorship gathered steam, young people from throughout the region (and beyond) swelled guerrilla

Coffee and Computers

Another example from El Salvador suggests something about the personal trajectories of these peasant intellectuals, as well as how the research process frequently challenged this author's preconceptions about campesinos.

In July 1994, I went to an unmarked building in a grimy, working-class neighborhood of San Salvador to interview René Hernández, a leader of the Society of Agrarian Reform Coffee Cooperatives (SOCRA). Hernández, born in 1957, had managed to complete fifth grade in his hometown of Candelaria de Santa Ana. A beneficiary of the first stage of El Salvador's agrarian reform, he belongs to a cooperative founded in 1980 that has twenty-five *manzanas* (17.5 hectares) of coffee. In the early 1980s, he attended courses on cooperative administration at CENCAP, a government agency; by the late 1980s, he was a leader of FESACO-RASAL and CONFRAS, aboveground cooperative organizations that were nonetheless influenced by one of the tendencies of the armed opposition FMLN (Goitia 1994, 181). As a representative of Salvadoran cooperativism, he has traveled to Germany, France, Israel, Mexico, Puerto Rico, and the rest of Central America. Since 1990, he has been on the board of directors of the state agricultural development bank and a member of its credit commission, a post he received as part of a deal between peasant organizations and the then minister of agriculture. "This," he says, "has been like a university degree [*carrera*] for me." And indeed, during our conversation, Hernández—sometimes jumping up to scribble on a white board with a marker—gave me a complex lecture about rediscount policies and interest rate spreads, value-added taxes, banks' loan portfolios, and government privatization policies.

On the way out, he nodded toward a room down the hall with some computer equipment and asked in Salvadoran, "*¿Querés ver el volado?*" (Do you want to see the thing-a-ma-jig?). Peasants with computers weren't a novelty for me anymore, and I wasn't interested in seeing yet another demonstration of some new spreadsheet program, word processing, or e-mail, but to be polite I responded, "*Va' pues*" (Okay). He nudged the mouse, and a screen full of columns of constantly changing numbers appeared. The "Best Investments" modem next to the 486-66 IBM-compatible had a cable running to a huge parabolic antenna on the roof. Hernández had hooked into the New York coffee market and was looking at up-to-the-minute price shifts and futures options. Grabbing the mouse, he started to open up windows with graphs of seven-, thirty-, and ninety-day price trends. "You see," he remarked with a sly smile, "now they can't lie to us about the price anymore."

ranks or collaborated from the Honduran and Costa Rican rearguards. With the triumph of the Sandinista National Liberation Front (FSLN), numerous "internationalists" (many of them political exiles) obtained positions in the government, party, news media, and pro-Sandinista research institutes and mass organizations. With the escalation in 1980 of armed conflicts in El Salvador and Guatemala and renewed warfare in Nicaragua beginning in 1981–82, hundreds of thousands of refugees—the majority peasants—fled their homes, seeking safety abroad and often elsewhere on the isthmus.

These movements of people brought participants in various kinds of peasant organizations into contact with one another. Members of Guatemala's Committee of Peasant Unity (CUC), exiled in Costa Rica, sought contacts with Costa Rican campesino organizations. But these ties were sporadic and focused largely on political solidarity and support rather than the problems of small producers. When Nicaraguan refugees began to pour across Costa Rica's northern border, members of Costa Rican campesino organizations who had backed the Sandinistas began to develop doubts about the revolutionary government's sometimes arbitrary land confiscation and the indiscriminate violence directed at rural communities suspected of harboring contras. Leaders of agricultural cooperatives and representatives of rural workers' unions from throughout the isthmus met occasionally in events sponsored by their regional organizations (CCC-CA and COCENTRA), both of which also had links to national and local organizations that represented campesinos outside the cooperative or union sectors.[7]

The internationalism of Nicaragua, promoted with the resources of a revolutionary party and state, provided an impetus for more frequent encounters. In the polarized atmosphere of Central America in the mid-1980s, revolutionary movements and campesino activists alike increasingly saw friends and allies elsewhere in the region as crucial for political success and even physical survival. The main clearinghouse for these contacts was the Nicaraguan National Union of Agricultural and Livestock Producers (UNAG). Founded in 1981 by smallholding peasants, cooperative members, and medium-size landowners who did not feel represented by the Sandinista-dominated rural workers' union (ATC), UNAG, despite its status as a mass organization, had a sometimes rocky relationship with the FSLN.[8] Leaders of Left-leaning Costa Rican and Honduran organizations passed through UNAG offices and toured rural cooperatives and commercialization projects, but these visits remained at the level of exchanges of experiences. Salvadoran and Guatemalan leaders also called, but at home they were often living clandestinely and had more urgent concerns than thinking in detail about the shape of their postwar agricultural sectors.

In addition to receiving visitors from outside Nicaragua, UNAG participated in two programs that accelerated the process of contact between peasant leaders from different countries. The first was a technology transfer program—*Campesino-a-Campesino* (farmer-to-farmer)—that trained peasant extensionists in sustainable cultivation practices and then had them provide technical assistance in and around their communities. Representatives of organizations from elsewhere in Central America at times participated in program workshops and events. Mexican agronomists who worked in UNAG also fostered a second program of technological exchanges between Mexican and Central American peasants. Campesinos in Vicente Guerrero, Tlaxcala, had a long and relatively successful experience with soil and water conservation techniques for hillside agriculture, methods that were sorely needed and little developed in Central America. Between 1987 and 1989, a Mexican NGO (SEDEPAC) sponsored several visits by Central American peasants to Tlaxcala.[9] "The Mexicans," recollected UNAG member Sinforiano Cáceres,

> helped us systematize our knowledge. How to make organic fertilizer. We knew that already, but they helped us to perfect it, giving us the quantities of each component. How to make live fences, windbreaks, dikes, how to convert an ox yoke so that a mule or horse could pull it. And there we met Guatemalans, Ticos [Costa Ricans], Panamanians, Central Americans. . . . From the Hondurans we learned about [nitrogen fixing] velvet beans. From the Ticos we learned more about crop rotation. And they logically learned something from us too."[10]

The European Connection

In order to understand how these expanding but still intermittent contacts gave rise to a Central America–wide association of peasant organizations, it is necessary to examine European policies toward the region. In the early 1980s, European governments looked on with growing alarm as the Reagan administration tried to topple the Sandinistas in Nicaragua and roll back the revolutionary movements in El Salvador and Guatemala. This apprehension—based on both fears of a major regional war and an analysis that stressed inequality and injustice rather than international communism as causes of the conflicts—led to extensive European backing for the Contadora peace process, initiated in 1983 by Mexico, Colombia, Venezuela, and Panama, and later for the efforts of Costa Rican President Oscar Arias that culminated in the 1987 Esquipulas Peace Accords.

In 1983–84, as part of Contadora, European governments provided funds to the Latin American Economic System (SELA), the consultative body of Latin American economics ministers, to set up the Support Committee for the Economic and Social Development of Central America (CADESCA). Headquartered in Panama, CADESCA became a channel for funding peace-oriented initiatives that the few other existing regional bodies could not easily handle and an alternative to the United States' near monopoly on aid to the region.[11] Initially, its programs focused on microenterprises, energy, environment, and regional economic integration. But as its director, Guatemalan economist Eduardo Stein, recognized, "there was a political aim in our technical efforts, which was maintaining places where dialogue could take place among Central Americans.[12]

Within a few years of its formation, CADESCA, at the request of the Central American ministries of agriculture and planning, started a major research program on food security issues (PSA). The ministers' concern grew out of a widespread recognition that the region—dependent on imports for over one-fifth of its cereals consumption (Arias 1989, 67)—was highly vulnerable to changing world market conditions and was threatened by free-market policies that discouraged basic grains production. Funded by the European Union (EU), the PSA also reflected a particular European critique of free trade in agricultural commodities that was one of the main sticking points in the Uruguay round of GATT negotiations (Santos 1988, 642–44).

Apart from organizing some national-level seminars with peasant leaders, the PSA worked largely with governments and had little to do with agricultural producers. It produced a series of technical studies that demonstrated the importance of smallholding grain producers in maintaining food security and food sovereignty, as well as a macroeconomic model that—in contrast to most mainstream models—was concerned primarily with measuring the impact of particular adjustment policies on a broad range of income and sectoral groups (Arias 1989; Arias, Jované, and Ng 1993; CADESCA 1990; Calderón and San Sebastian 1991; Dévé 1989; Martínez 1990; Torres and Alvarado 1990).

As the PSA wound down in 1990, CADESCA started a food security education program (PFSA) intended to make the PSA's findings available to government functionaries, who would then be better able to formulate policy, and to peasant leaders, who presumably could then participate in the food security debate and influence policymakers. This concern with training leaders of popular organizations reflected a view of democratization shared by CADESCA and the Europeans that stressed the participation of civil society in policymaking processes, a conception that contrasted with the U.S. emphasis on free elections, legal reforms, and formal institutions (Cohen and Arato 1992).

Directed by Salvador Arias, a European-trained economist and former minister of agriculture in El Salvador,[13] the PFSA hired consultants to direct key program areas (*ejes*): credit, marketing, land reform, technology, and environment. Most of the consultants were from outside Panama and also served as liaisons with campesino organizations in their countries. Generally economists and sociologists with considerable field experience and at least some graduate training, they became strategic figures in the articulation of program objectives and in identifying which organizations to invite to regional seminars.[14]

Although campesinos who participated in the PFSA differ in their opinions as to the ultimate usefulness of the program, they coincide in describing the seminars as having important side effects, at least to some degree unintended. "In the first meeting," Sinforiano Cáceres recalled,

> we discussed our problems and found that many were the same, that we had more in common than we had differences. The "vest cut" [*corte de chaleco*] that structural adjustment had done on us left us all in the same conditions. . . . [15] They screw us in different ways, but in the end they're the same. . . . The IRA in El Salvador, ENABAS in Nicaragua, the CNP in Costa Rica [state commodities boards] all now play the same role: cheap food for consumers, low prices for us. . . . The "agricultural modernization law" in Honduras is the same as the plan to destroy the land reform projects [*asentamientos campesinos*] in Panama. . . . And [in Nicaragua], through the market, a process of agrarian counterreform is also taking place.[16]

In the first regional PFSA seminar in November 1990, several organization leaders demanded, as a condition of their participation, that the program provide extra time so that peasant groups from different countries could develop their own agenda and discuss common problems. This, as well as the overly academic tone of program documents and specialists' presentations, at times reportedly caused friction and misunderstandings between peasant leaders and CADESCA. But as one rural sociologist present at this first encounter pointed out:

> Once the leaders met and got to know each other, many for the first time, and they realize they've all always been more or less in the same situation. . . . Well, as always in these things, the whole is more than the sum of the parts and the people began to elaborate their own agenda. The program's agenda had another rhythm. Our agenda turned out to be too rigid for the needs and expectations of a movement that was beginning to find its identity. We had to reorganize the program several times and adjust it and adjust it. They were telling us what themes to cover, what things they wanted to know more about. But this turned out to be precisely the virtue of the program. It wasn't easy, because when you're managing a program, you're the contracted technical personnel, you understand that things have to follow a certain schedule. But that doesn't always coincide with the rhythms of the people. So there were a lot of difficulties. But the vision that prevailed was to take a chance on them, to accommodate their process.[17]

By the end of the second PFSA seminar in February 1991, peasant organization representatives had formed a provisional commission, coordinated by

Reading Chinese

According to Carlos Hernández, "many of the documents were too technical. I, having read a bit, might be able to understand them, but for other *compañeros* it was as if they were given a document in Chinese." These comments were echoed in many other conversations with PFSA participants. I acquired a sizable box of PFSA documents in 1991 from a highly articulate activist in northern Costa Rica, a voracious reader who had come close to graduating from high school. While packing his belongings prior to moving, he threw up his hands and exclaimed, "If these things interest you, take them! I'll probably never read them."

For PFSA director Salvador Arias, however, "this was a conscious thing on our part. At times there is an oversimplification of the training given campesino leaders. It is almost reduced to ABCs. We didn't agree with that. We said we'll give them complicated topics and we'll explain them, so that they raise their level. . . . Some resisted, but in the end it was positive, because the campesino leadership began to have a new capacity, a new vocabulary, a new use of social and economic categories. They pressured us to write things in a certain language. But we said, 'No, we're not going to do that. I can explain globalization in the simplest way and you will understand me. But if you don't handle the terminology used by the politicians with whom you're negotiating, even if you know about globalization, you're not going to understand them, because they aren't going to use your categories. You have to use their words. When you're negotiating, you can't ask ministers to negotiate at your level. You have to raise the level.' And now it's easy to find campesino leaders in Central America who can speak about macroeconomics, about economic adjustment."

the Costa Rican delegation, with a view to forming a Central America–wide association (ASOCODE 1991b, 4). The process took on a certain urgency because of the Central American presidents' plan to hold a summit in mid-1991 where major decisions would be made about regional agricultural trade policies. The technical support of PFSA specialists and the prospect of continued European funding, through CADESCA and other agencies, clearly conditioned the pace and style of organization as well.

In April, the provisional commission sent a lengthy letter to the Central American presidents on the eve of their summit. It opened with a condemnation of "economic structural adjustment, which even the international financial institutions recognise directly attacks the interests of the majorities of our peoples." It called on the presidents "to promote the ongoing processes of political opening and reconciliation [*concertación*]" and reminded them "that we have already elaborated alternative and integral development proposals, which we believe are possible to execute," such as "vertically integrated production, which will permit us to break out of our historical situation of only producing raw materials and to obtain the benefits that are generated by the agro-industrial processing of our production." Finally, the letter cautioned, "if our rights are not respected, the process of peace, so precarious and difficult to reach, will escape from our hands and then, with the deepening of our poverty [*miseria*] and marginality, social confrontation and war will continue, frustrating our peoples' desire to live in harmony, in a stable and peaceful social climate, with justice and real democracy" (Consejo Nacional 1991, 1).

A document by this commission for the summit addressed structural adjustment, as well as relations with NGOs and international "cooperation" agencies (Comisión Centroamericana 1991). Notably, it employed a novel rhetorical strategy—later common in peasant movement declarations—of appropriating discourses of incontrovertible legitimacy. It chose such discourse as the Latin American bishops' 1979 condemnation at Puebla of "economic, social and political structures [that cause] inhuman poverty," the Central American presidents' own call at Esquipulas for "egalitarian societies, free of misery," and UN economists' notions of "economic adjustment with a human face" (Bustelo et al. 1987).

The statement's authors declared that since beginning "the slow process of regional coordination in 1988," they had manifested "mature and responsible attitudes," a phrase that in the Central American context could probably be understood as meaning that they had eschewed guerrilla violence. They stressed their independence from political parties and their commitment to struggling not just for conjunctural demands but also for obtaining a role in shaping long-term policies affecting the agricultural sector. They pointed out that in Costa Rica, Nicaragua, and Honduras they had sustained negotiations with ministers of agriculture and presidents. Finally, they noted that the international financial institutions and "the governments and dominant sectors in the different countries and the international solidarity cooperation agencies themselves" were beginning to operate at the Central American regional level and that campesino organizations now had to do the same if they were to influence policies that affected them (and, presumably, if they were to gain access to "cooperation" funds).

Nations in the Region

The creation of a Central America–wide association of peasant organizations grew out of shared problems, but it raised numerous issues related to national particularities. Most importantly, the political situations in the different countries varied greatly, from relative openness in Costa Rica to severe, continuing repression in Guatemala. Economic stabilization and structural adjustment, which began in Costa Rica in 1983, were just beginning in Honduras and El Salvador. Nicaragua and El Salvador were emerging from wars and Panama from the U.S. invasion. Belize—largely English speaking—had always related to the Caribbean more than to Central America. Honduras had the oldest and largest peasant movement in the region, and Panama—with its canal- and service-based economy—had neither a numerically large peasantry nor strong campesino organizations.[18] Peasant leaders, of course, had different backgrounds, constituencies, aspirations, political loyalties, and levels of sophistication.

From the beginning, the Costa Ricans and Nicaraguans had played key roles, although for rather distinct reasons in each case. The Nicaraguans in UNAG had the largest and most consolidated peasant organization in the region. They had ties to a revolutionary party that, at the beginning of the 1990s, appeared to have possibilities of returning to power, and early on, they had taken the initiative in meeting with organizations from elsewhere in the isthmus. At least some Nicaraguan leaders apparently believed that UNAG, by virtue of its size and position, ought to dominate any Central America–wide association.

The Costa Rican peasant movement was smaller and politically more heterogeneous; it consisted of several large but resolutely apolitical cooperative-sector organizations, some independent local groups, a centrist small producers union (UPANACIONAL) based in highland coffee and vegetable growing zones, and a Left-leaning coalition called Justice and Development (CCJYD), which included a diverse collection of small organizations and cooperatives. In 1991, UPANACIONAL and CCJYD, previously distant from each other, united in a coordinating body (CNA) to carry out joint negotiations with the government (Román 1994, 79).[19] The Costa Rican organizations had the longest experience with and the most developed analysis of economic structural adjustment programs (SAPs).

In particular, some leaders of CCJYD, which in the mid-1980s had taken a belligerent stance against Costa Rican neo-liberalism, felt that they were in a privileged position to foretell what would befall campesinos in other countries where SAPs were just beginning.[20] In 1988, before the PFSA seized the initiative for organizing regional meetings, the Costa Rican organizations had already formed a short-lived three-person committee to seek EU funds for a gathering of peasants from El Salvador, Nicaragua, and Honduras (Hernández Cascante 1992, 1). Despite their concern for organizing at the regional level, however, the very sophistication of their analyses meant that they were not entirely able to transcend negative stereotypes of Costa Ricans held by other Central Americans.[21]

The unanticipated sophistication of the Costa Rican leaders had earlier impressed PSA specialists. In their first meeting in 1988, "the national technicians who worked with the PSA–Costa Rica were surprised by the peasants' arguments [*planteamientos*]. It was not common to hear campesino proposals, and it was even stranger [*más singular*] to find that those proposals constituted a broad alternative to what the government's economic team was then negotiating with the World Bank and the IMF" (Hernández Cascante 1992, 1–2).

Not all countries' participants in the emerging Central America–wide campesino association rejected relations with political parties. The Costa Ricans had years earlier broken with the organized Left (Edelman 1991), and the Nicaraguan UNAG had, after the Sandinistas' 1990 electoral defeat, declared its autonomy from the FSLN.[22] The many Honduran organizations did not, for the most part, have "organic" ties with political parties (though they constantly cut deals and formed conjunctural alliances with them). In El Salvador, however, peasant groups of Left, Right, and Center had close links to parties. This was a legacy of the 1980–91 civil war, when each of the five parties in the FMLN guerrilla coalition, as well as the right-wing and centrist parties, sponsored parallel union, peasant, and youth organizations. Salvadoran participants in the PFSA and subsequent regional campesino meetings claimed that they separated union [*gremial*] and party loyalties, but—contradictorily—they also remained proud members of their respective party groups.[23] This led to charges of verticalism from other countries' representatives, who disliked Leninist-style party discipline and sectarian work styles (Biekart and Jelsma 1994, 10; Hernández Cascante 1992, 3; 1994, 252).

Belize and Guatemala did not participate in the PFSA and were secondary players in the emerging campesino association. Belize, long identified culturally, linguistically, and politically with the English-speaking Caribbean, was little known to PFSA organizers and had few farmers organizations; the Belize government was relatively uninterested in Central American integration, since it was already in the Caribbean Common Market (CARICOM).[24] Guatemala remained on the sidelines for different reasons. Its largest organization, CUC, had links to the armed Left and still operated in partial clandestinity. Both CUC and many smaller and less militant organizations were frequent targets of brutal repression. Their leaders accorded greater priority to physical survival and to the struggle within Guatemala than to establishing high-profile links with their counterparts in neighboring countries.[25]

Organization of ASOCODE

In Tegucigalpa, in July 1991, the First Regional Campesino Conference, with delegates from throughout the isthmus, agreed to found the Association of Central American Peasant Organizations for Cooperation and Development (ASOCODE). The conference approved a position paper for the region's agriculture ministers and the tenth Central American presidents' summit, which took place later that month in San Salvador. This statement—termed ASOCODE's (1991a) productive strategy—affirmed that small producers make rational and intensive use of their scarce resources, but that they were nonetheless threatened with extinction. It condemned SAPs, skewed landownership patterns, the reversal of agrarian reforms, and the hypocritical protectionism of the countries providing the growing quantities of surplus food aid that undermined peasant grain producers. Finally, the paper called for preferential fiscal, credit, and pricing policies for small producers; for the participation of peasant organizations in agricultural-sector policy-making bodies and state development banks; for campesino organizations to be given first purchase options for public-sector agroindustries undergoing privatization; for free trade in grain within the region, but with protection against highly subsidized non–Central American producers; and for improving the governments' capacity for evaluating and controlling imported technologies, especially biotechnologies.[26]

The summit produced the Action Plan for Central American Agriculture (PAC), which instructed the ministers of agriculture to develop data on the number of producers, production costs, and production and productivity of each key crop. PAC included several measures to liberalize intraregional trade, most importantly, a rapid reduction in state involvement in commercializing agricultural products and a system of uniform regional price bands for basic grains (Presidentes Centroamericanos 1991).[27] To the surprise of many, the presidents' final summit declaration resolved "to receive with special interest the proposals of the Association of Central American Peasant Organisations for Co-operation and Development and to instruct the appropriate [government] institutions to consider and analyse them in order to find adequate responses to the issues they raise" (quoted in ASOCODE 1991b, 23). This gesture was, of course, largely rhetorical, but it represented a degree of recognition as a legitimate political force that few of the campesino activists had expected.

In December 1991, campesino organizations from throughout Central America came to Managua for ASOCODE's founding congress, an event that mixed resolutions, speeches, and association business with "a rich and lively flow of sentiment, denunciation and synthesis from the singers, poets and musicians" in the different countries' delegations (ASOCODE 1992a).[28] The congress formalized a coordinating commission made up of two delegates from each participating national coalition.[29] It also elected a general coordinator, Wilson Campos Cerdas, a charismatic thirty-two-year-old Costa Rican who, since 1988, had played a key role in organizing and securing funds for regional meetings.[30] Largely at the insistence of the Nicaraguan delegation, the congress decided that if the coordinator were from Costa Rica, the association's headquarters would have to be in another country. In an effort to balance tensions between the Costa Ricans and Nicaraguans, a UNAG functionary was elected vice coordinator, and Nicaragua was chosen as ASOCODE's seat.

The new association conceived itself not as a new supranational bureaucracy but as a "meeting table" (*mesa de encuentro*) for the national coalitions, where decisions would be made by consensus,[31] and as a lobby that would struggle to defend campesino interests in the international, regional, and national arenas. The congress specified that these interests included:

1. "Guaranteeing small and medium-size producers access to land, credit and technical assistance, as well as processing and marketing of their production."

2. "Assuring respect for small and medium-size producers' cultural roots, so that the development of Central American societies will be compatible with their idiosyncrasies and way of life."

3. "Achieving full recognition and participation in political and economic decisions at the national, regional and international levels."

4. "Working for a true peace and true respect for the elemental human rights of small and medium-size producers."

5. "Promoting conservation of Central America's ecological systems" (ASOCODE 1991b, 25).

The assembled delegates resolved to request an invitation to the upcoming Earth Summit in Rio de Janeiro, to demand that the Central American presidents and agriculture ministers enter into negotiations about free trade in basic grains, and to convoke a forum with international solidarity cooperation agencies. Dutch organizations had already invited ASOCODE to attend a planning meeting in Belgium about how to lobby at the upcoming San José VIII meeting of European and Central American foreign ministers (ASOCODE 1991b, 7–9).

Alternative Messages and Funding the Messengers

The program was ambitious, the enthusiasm and momentum great, and the political conjuncture—within and outside the region—apparently propitious.

But how was this challenging agenda to be put into practice and funded? In 1992, two delegations toured Europe; the first group, whose trip was coordinated by the Amsterdam-based Trans-National Institute (TNI), received a welcome beyond all expectations. Representatives of the governments of Holland, Denmark, Sweden, Norway, Germany, Belgium, and France, as well as high-ranking EU parlamentarians and officials, met with the ASOCODE envoys, often "for more time than protocol usually requires for this kind of interview." An internal organization report noted that the government representatives "listened with curiosity and at times with surprise at the level of our arguments and our knowledge regarding global economic and agricultural issues and the political, economic and social problems of our region. In sum, the result is highly favourable to ASOCODE." The tour also reinforced links with a variety of European NGOs and foundations, university groups, news media, alternative commerce campaigns, and farmers' organizations (which, the report declared unself-consciously, "helped us become aware of the backwardness that we have") (ASOCODE 1992b; see also Chapter 5 of this book).[32]

How successful were ASOCODE's European tours and related efforts in securing material support? By the end of 1992, the association inaugurated its Managua headquarters, a spacious rented house in an upper-middle-class neighborhood one block away from the home of President Violeta Barrios de Chamorro. Physically, the office resembles that of any large Central American NGO, with computers and copying equipment, offices for professional staff, secretaries, a guard, a maid, and a driver with a four-wheel-drive vehicle. By 1993, the organization had an annual budget of over US$300,000; a monthly subsidy of US$1,000 was allocated for each of the seven national coalitions (ASOCODE 1993b, 18). By 1995, this subvention had risen to between $4,000 and $5,000 per month for each coalition.[33] The general coordinator's salary in 1993 was US$13,200 (not including the end-of-the-year bonus), a handsome income for a midlevel professional in Central America. The largest budget line—apart from personnel, overhead, and the subsidies for national organizations—was for travel to planning meetings, most of which involved bringing people from six other countries (ASOCODE 1993b, 18).

This substantial flow of resources permitted ASOCODE to sponsor frequent seminars with campesino leaders devoted to credit, marketing opportunities, agricultural and agroforestry technology, administrative and lobbying skills, and other needs. The association also produced a constant flow of proposals and position papers and maintained a regular presence at regional intergovernmental meetings. In Panama, in December 1992, after intense ASOCODE lobbying, the Central American presidents' summit issued what became known as the Agricultural Commitment of Panama

(CAP), a series of guidelines to orient regional policy. Although much of CAP called for eliminating remaining barriers to free trade, food security was also an important concern throughout the declaration.

Several points suggested that campesino lobbyists had achieved some impact. Among other things, the presidents called for protecting small basic grains producers from fluctuations and distortions in international markets; for creating a regional fund for improving smallholders' access to technology, credit, and processing facilities; and for incorporating representatives of the public and private agricultural sectors into policymaking processes and international commercial negotiations (Presidentes Centroamericanos 1992). Even at the moment of conception, it was clear that many of CAP's promises would likely go unfulfilled. Nevertheless, they constituted a significant reference point for future dialogue and negotiations. The presidents would point to CAP as evidence of their good intentions, and the peasant organizations in ASOCODE would demand fulfilment of CAP provisions favored by their constituencies.

Despite the apparent concessions in CAP, some of the presidents were far from pleased at the persistent presence of peasant lobbyists at their regional meetings. Especially since the 1991 confrontation between peasant leaders and government ministers over grain price bands policies (see note 27), the more conservative governments had viewed first the PFSA and then ASOCODE with growing consternation. The Callejas government in Honduras took the perceived threat seriously enough to bring right-wing peasant leaders to the Panama summit and then to employ a classic Honduran tactic for dividing popular movements: the creation of a parallel organization.[34]

ASOCODE rapidly succeeded in gaining regional and international political recognition and legitimacy, though at times this caused tensions with the participating national coalitions. In December 1993, for example, the association held its second congress in Guatemala and invited President Ramiro de León Carpio to address the gathering. This was a calculated effort to provide a shield for CONAMPRO, ASOCODE's organization in Guatemala, and for ASOCODE representatives traveling there, who on earlier occasions had suffered harassment at the airport.[35] CONAMPRO representatives wanted to use the congress and de León's presence to raise pressing issues of "massacres, . . . forced [military] recruitment, . . . and political persecution."[36] But ASOCODE leaders, concerned about possibly offending the president, exerted pressure to remove all such references from CONAMPRO's inaugural statement to the congress. The final version of the CONAMPRO coordinator's speech made only vague allusions to war and repression and to Guatemala's "long and dark night, which has no end in sight" (ASOCODE 1993a, 7).

For ASOCODE, the strategy of lobbying ministers and presidents had several obvious strong points:

- It buffered the national organizations against repression.

- It served as an important source of information about impending policy shifts at the national and regional levels.

- It demonstrated to international organizations and funders that the peasant movement was not inveterately confrontational and that it was capable of offering alternative development proposals and was willing to negotiate with policymakers.

- It contributed to democratization inasmuch as peasants and other sectors of civil society had gained the right (and access to forums) to express their demands and to insist on compliance with government commitments.

- In several countries, pressure from ASOCODE contributed to winning national organizations' demands for participation in policymaking bodies, such as public-sector agrarian banks and bipartite agricultural sector commissions composed of ministerial and peasant organization representatives.

Successes in lobbying and negotiations depended significantly on the peasants' growing capacity for appropriating and reshaping official discourses—and not just the presidents' frequent but vague calls for *concertación* (consensus and reconciliation) or for the participation of civil society. The specificity of this approach is suggested in the comments of a Salvadoran activist:

> To speak of the development of El Salvador is to speak of the Lempa River basin: half the country, 10,000 square kilometers, the source of 98 percent of our energy, our principal water source. The country's future has to do with the Lempa basin. . . . We use this to make the traditional demands of the peasant movement: land titling, credit, marketing, technical assistance. But we negotiate around what most interests the country: energy and water. Who lives on the slopes of the Lempa basin? Poor peasants producing basic grains without technology or assistance; they can't change their relation to the land because their rights to it haven't been recognized. Even the U.S. Agriculture Department recognizes that if people don't own their land it's difficult to change their relation with natural resources. So in negotiations we raise the banner of the Lempa basin. "You're interested in energy? We don't even have energy, we don't have light. You're interested in preventing sedimentation of the dams? You invest millions of dollars in dredging. But if you want to *prevent* runoff and sedimentation, we have to conserve the soils, and only the peasants can do that." This argument is like a new weapon for negotiation.[37]

Peasant negotiators had become adroit at appropriating and reshaping official rhetoric. Their success in lobbying and negotiations, however, also depended on the willingness of those in positions of power to compromise. By mid-1994, the members of ASOCODE's coordinating commission agreed that the region's governments had demonstrated lack of political will; the promises of CAP—and many others—had not been kept. Wilson Campos summed up the mood, saying, "We've forced them to recognise us as a legitimate force. But now, after two years, we've been in four summits and over twenty regional forums. We're seeing that they've made a lot of promises that haven't been kept" (quoted in Edelman 1994, 31; see also ASOCODE 1994a).

Government intransigence brought increasing calls from within ASOCODE for a return to traditional pressure tactics—marches or even street or highway blockades or building occupations. Most agreed, however, that any demonstration should be carried out simultaneously in all seven countries and without abandoning efforts to affect policy through other methods and at other levels. Even before any show of force, the threat of action won some concessions. As a coordinating commission report from September 1994 indicated:

> In November, for the first time, we will have an [entire] day to work with the Central American Agriculture Ministers; but we only obtained this one-day audience because we sent a letter saying that ASOCODE was considering the possibility of regional pressure and their response was to immediately give us that working day. (ASOCODE 1994b, 5)

The possibility of regional pressure had, in fact, already been considered and decided upon. On 10 October, organizations in the five of the seven countries staged simultaneous marches for a variety of specific national demands and in protest against the governments' unwillingness to consider modifications in national structural adjustment. The large turnouts—especially in Honduras and Costa Rica, where the presidents received delegations of demonstrators—constituted a significant show of strength and were a useful morale builder for the organizations. But the marches appeared to do little to break the impasse between the governments and the peasant organizations.[38] Two weeks later in Honduras, Wilson Campos addressed a major international meeting on peace and development in Central America, but he did so not as ASOCODE's coordinator but as a leader of ICIC, a broader civil society coalition (Conferencia Internacional 1994, 71–79). At the more important Central American presidents' summit in March 1995 in El Salvador, ASOCODE did not receive the by now customary invitation to address the meeting, despite the event's focus on social welfare issues.[39]

Increasingly, the association turned inward, continuing to seek access to presidents and ministers, but working more to strengthen the national coalitions and alliances with nonpeasant organizations, to identify whatever opportunities might arise as part of the free-market transition, and to foster campesino entrepreneurial and administrative capacities (see ASOCODE 1994c, 1995). In Honduras, El Salvador, and Guatemala, organizations linked to ASOCODE staged major land occupations in 1995 and early 1996, producing occasional concessions but also new victims of state repression.[40]

Problems and Prospects of a
Transnational Peasant Movement

The rise of a transnational peasant movement in Central America does not easily fit existing theories about social movements, in part because social scientific debate about collective action frequently takes place at such a high level of abstraction that it lacks concrete political content.

As a campesino movement, ASOCODE has broken with the local, purely agrarian protest orientation that historically characterized so many peasant mobilizations in Central America and elsewhere. At the same time, it rejects the strategy of peasant wars, which consumed so many of its supporters during the 1980s. Campesino involvement in high-level lobbying, in international networks and alliance building with nonagricultural sector groups, and in the elaboration of detailed and often sophisticated development proposals marks a new stage in what is both a very old and a very new social movement (see Warman 1988).

ASOCODE and its member organizations share a number of features of the new social movements, or NSMs:

1. They shun party politics and the verticalism of what remains of the traditional Left (with the exception, perhaps, of the Salvadorans, who are largely FMLN veterans).

2. Like the NSMs, they have forged ties to and participate in environmentalist, women's, and indigenous groups that reflect concerns beyond the narrowly economic or agrarian demands of the traditional peasant movement.

3. They have a significant identity-based dimension that involves struggling in the face of threatening economic changes for the right to continue being peasants—for the campesino idiosyncrasy, as they sometimes describe it.

At the same time, key aspects of contemporary Central American peasant organizing suggest both the oft-cited Eurocentric bias of most NSM theory and the campesino organizations' roots in age-old agrarian conflicts. Most importantly, the classlessness and the centrality of cultural over material struggles, said to be typical of NSMs (Escobar 1992; Jelin 1990; Melucci 1989; Olofsson 1988), are completely alien to activists who are acutely aware of the worsening economic polarization of Central American societies and of the ways that international financial institutions affect their livelihood.

Transnational peasant organizing not only emerged from serious material crises in the rural sector, but its impact depended significantly on resources provided by foreign cooperation. The notion propounded by some NSM theorists (Escobar 1995) that the Third World is entering a postdevelopment era rings hollow in contemporary Central America. Certainly the rise of regional peasant organizations is part of a spirited critique of mainstream development practice—of a struggle over how development is to be defined and carried out—but it is also indicative of a profound longing for the levels of well-being conventionally associated with development. If nothing else, the fact that peasants at the Central American and national levels put the word *desarrollo* in their organizations' names suggests both that development remains an important aspiration and that they are willing to appropriate, contest, and refashion its meaning.

Several concerns raised by resource mobilization approaches—the main alternative paradigm to NSMs in the social movement literature—are clearly relevant for understanding the internationalization of Central American peasant movements. This framework's emphasis on strategies of recruitment and organization formation and on activists' (or movement entrepreneurs) career trajectories constitutes a useful complement to identity-oriented NSM theory. In the case of the peasant organizations that united in ASOCODE, European backing was (and is) clearly key, even if threats to a cherished smallholder identity often provided a goad to action at the individual or national level and even if—without foreign cooperation—something resembling ASOCODE might have emerged anyway with a smaller budget and a lower profile ("taking buses to Guatemala or Honduras, the way *we* do, rather than airplanes the way *they* do," as a mildly critical participant in the Central America–wide labor union group COCENTRA expressed it).

The availability of European (and other) funds raises questions about the mix of motives of those participating in and leading national and international peasant organizations, the long-term possibilities of movements vulnerable to the growing fiscal conservatism of European societies, and the ultimate political and economic impact of the internationalization process. ASOCODE leaders assert that "no governmental or non-governmental organisation has the right to represent itself as the parent or creator [*padre o gestor*] of this process" (1992a, 4).[41] Nonetheless, solidarity cooperation has at the very least facilitated a new kind of peasant political practice.

Connected to this is the emergence of a new kind of peasant politician with a new kind of identity, though not one of the idealized identities romanticized by some NSM theorists. Being a *dirigente*, or leader, has become a career path, with the security of a monthly salary and possibilities for advancement and foreign travel.[42] As ASOCODE's economic support for the national coalitions increases, the number of activists directly or indirectly on the payroll has grown. Even when these cadres conduct themselves with the utmost integrity, as generally appears to be the case, the *perception* that they form a privileged group almost inevitably causes frictions.[43] Those outside the top leadership now sometimes mutter about *yuppis campesinos*, *el jet set campesino*, or *la cúpula de cúpulas*. In traditional Central American peasant politics, receiving a salaried position can be a payoff or a cause for envy. It may have an "odor of corruption," even when the employer is not a government but a popular organization, and even when the newly fortunate employee is scrupulously honest.

In some respects, the dilemmas of a well-funded peasant organization mirror those facing NGOs—and at least some peasant leaders fear that ASOCODE could become an NGO, an intermediary that seeks money for peasant groups, rather than a popular organization that represents them politically (Candanedo and Madrigal 1994, 164). Broadly speaking, scholars and development practitioners make two competing sets of claims about NGOs. Some take a basically positive view of the rapid growth in the numbers and prominence of NGOs, emphasizing that this represents the emergence of a strengthened civil society in states undergoing transitions to democracy and economic retrenchment (Gordon-Drabek 1987). Others, more critical of the NGO boom, maintain that donor NGOs and development bureaucracies form local NGO "satellites" to justify their existence, and that they export and represent as autochthonous (*criollo*) their own ideologies of development. According to this argument, articulated by both development specialists and peasant activists (Kruijt 1992; Lofredo 1991; O'Kane 1992), the professionals displaced from contracting public sectors who become NGO personnel constitute a new upwardly mobile social group with vested interests in the growth of NGO-supported projects and, ultimately, in continuing dependent relationships with the communities they serve. The tension between the NGO–as–germ of civil society and the NGO–as–career path is not only found in NGOs, an irony not lost on some peasant activists.

A significant contest over representation lies behind both the rise of ASOCODE and the fears of some of its supporters that it could become a NGO. In the late 1980s and early 1990s, peasant organizations in Central America often charged that NGOs were a parasitic intermediary stratum that wasted resources on fancy offices and vehicles and pretended to speak for popular sectors as a means of courting donors (Edelman 1991; O'Kane 1992). One effect of this vehement, though frequently justified, offensive was that various developed-country governments and donor NGOs withdrew support from Central American NGOs and began to fund peasant organizations directly. Other Central American NGOs repudiated or muted earlier claims to represent the peasantry and reorganized themselves as unassuming providers of technical, administrative, or research expertise willing to serve peasant organizations or cooperatives on a contractual basis.

Once the peasant organizations felt more secure economically and that a more balanced and respectful relationship had been established with the NGOs, relations between the two groups improved; at the regional level, development NGOs formed a network (CCOD) that became an ally of ASOCODE. In some cases, peasant leaders played key roles in setting up new NGOs to carry out technical and policy studies for their movement.[44] In effect, the campesino movement had been successful in telling foreign donors, "don't fund *them*, fund *us*."

During its short existence, ASOCODE has succeeded in transcending several of the historical sources of weakness and division in peasant movements. Its rejection of political party ties, its genuine ideological pluralism, and its commitment to internal and external dialogue and consensus building have permitted it to coordinate a diverse group of organizations in different national settings and to achieve a remarkable degree of regional and international recognition. This ambitious project of coordination and action, however, perhaps inevitably brought a degree of separation from both national coalitions and base organizations. This hinders information flows in both directions and fuels the perception that the regional leadership constitutes a distant elite.

The association's concentration on high-level lobbying and organizational consolidation has meant that many of the ambitious alternative development plans hashed out in regional seminars have yet to be applied on the ground— "to land" or "come down to earth" (*aterrizar*), as the frequent fliers in the leadership put it. To some extent, the agenda of ASOCODE and its constituent organizations is now donor driven—a different and less onerous kind of conditionality than the political party domination of earlier peasant organizations, but one that nonetheless preoccupies some of its constituency. Much of the attention now given to gender, indigenous, and, to a lesser extent, environmental issues clearly derives primarily from European rather than Central American sensibilities (Candanedo and Madrigal 1994, 119).[45]

Perhaps the most encouraging aspect of these contradictions is that they are understood, debated, and addressed within ASOCODE with a frankness that has few antecedents in earlier peasant movement practice. Campesino organizations are political projects, not profit-generating enterprises, and many of these tensions are probably unavoidable (compare Landsberger and Hewitt 1970). In Central America, at least, even elite business lobbies rely heavily on foreign cooperation, usually from USAID (Rosa 1993; Sojo

1992). That the peasant movement has sought funds abroad could even be interpreted as one indication of growing realism, specialization or professionalization, and maturity. That ASOCODE has managed to bring together such a diverse and fractious collection of groups from seven different countries is, given Central American peasant movements' long history of factionalism, nothing short of remarkable.

Notes

I gratefully acknowledge research support from the U.S. National Science Foundation (grant #SBR-9319905), the U.S. National Endowment for the Humanities (#FA-32493), and the Wenner-Gren Foundation (#5627). I presented an earlier version of this chapter at the 1995 meeting of the Latin American Studies Association and at an October 1995 seminar of the Yale University Program in Agrarian Studies. Many participants in these forums made useful comments on the manuscript, but I am especially appreciative of the insightful and provocative remarks of Anthony Bebbington, Michael Dorsey, and James Scott. I am also indebted to León Arredondo, Mauricio Claudio, Alcira Forero-Peña, Néstor Hincapié, and Víctor Ortiz, who helped to transcribe taped interviews. This project benefited immensely from the friendship, trust, and logistical support provided by peasant activists, secretaries and *técnicos* in organization offices, and others too numerous to mention. A todos, muchas gracias.

1. "Region" and "regional" are used here to refer to Central America as a whole. Other Central America–wide peasant organizations include sector-specific groups such as the Union of Small and Medium Coffee Producers of Mexico, Central America, and the Caribbean (UPROCAFE), founded in 1989, and the Cooperative Confederation of the Caribbean and Central America (CCC-CA), founded in 1980, which has an agricultural co-op section.
2. Regional integration processes accelerated following the June 1990 Central American presidential summit in Antigua, Guatemala. In December 1991, the Tegucigalpa Protocol created the Central American Integration System (SICA), which incorporated the periodic regional meetings of presidents and ministers and the regional parliament (PARLACEN) founded as part of the 1987 Esquipulas Peace Accords. In contrast to the Central American Common Market of the 1960s, which relied on high extraregional tariffs to stimulate industry geared toward regional markets, current integration efforts are antiprotectionist and emphasize nontraditional agricultural exports (and *maquilas*, or garment assembly plants) as the engine of growth. The creation of SICA and of regional business lobbies (such as FEDIPRICAP) helps explain why peasant leaders felt the need to organize at the Central American level.
3. In 1989, negotiations for a new coffee agreement stalled, and world prices plummeted 40 percent in one month to the lowest levels in over twenty years (Pelupessy 1993, 39–40). Prices registered a modest rise in mid-1994, when frost struck coffee-growing regions of Brazil, but they again tumbled precipitously in mid-1995.
4. In 1960, 33.2 percent of Costa Rica's population was urban; by 1988, this had risen to 50.3 percent; the urban population growth rate was among the highest in Latin America in this period (IDB 1989, 458). During the 1970s, the postsec-

ondary educational system expanded dramatically; by 1980, a remarkable 27 percent of the university-student-age population was enrolled in institutions of higher learning (World Bank data cited in Mendiola 1988, 82).

5. This goal has sometimes conflicted, however, with organizations' need to mobilize politically, since to a certain extent the peasant-as-rustic remains a critical symbol for garnering sympathy from policymakers and the public (Edelman 1991).

6. In 1994, for example, regional cooperative, community, labor, NGO, and small enterprise and farmers' networks formed a lobbying group called the Civil Initiative for Central American Integration (ICIC).

7. Interviews with Wilson Campos, ASOCODE, Costa Rica, 11 June 1994, and Panama, 16 June 1994; Carlos Hernández, CCJYD, Costa Rica, 16 June 1994; José Adán Rivera, ATC, Nicaragua, 29 June 1994; Sinforiano Cáceres, FENA-COOP, Nicaragua, 4 July 1994.

8. Virtually all UNAG leaders belonged to the FSLN, and some held high positions. Nevertheless, as UNAG functionary Amílcar Navarro recalled, "at that time [c. 1981], to own means of production was to be bourgeois. It was thought that the peasant movement had the same interests as the workers' movement, as salaried agricultural workers, but that's not so. . . . The Sandinista Front supported the workers' movement much more than the peasant movement. The Front had intellectuals, students, workers—and very few campesinos. They didn't understand the campesino who wanted to make his land produce, to sell his products at a good price, to have technical assistance. . . . The workers struggled to work less, five hours instead of ten hours. But we're employers and we're paying these guys, so I can't support them when they say they want to work less" (interview, Managua, 1 July 1994).

9. Interview, Rubén Pasos, FUNDESCA, Managua, 6 July 1994.

10. Interview, Sinforiano Cáceres, Managua, 4 July 1994.

11. CADESCA was originally intended to be a short-term undertaking. In 1994, it ceased to exist as an intergovernmental entity and was replaced by a private foundation (FUNDESCA), funded primarily by the EU, European governments, and the Nordic-country NGOs participating in the Copenhagen Initiative for Central America (CIFCA). FUNDESCA took over CADESCA's Panama offices and carried on its existing projects.

12. Interview with Eduardo Stein, Panama, 22 June 1994. Stein became Guatemalan foreign minister in 1996.

13. Arias received a *licenciatura* in economics from the Jesuit-run Universidad Centroamericana in San Salvador in 1974. In 1975–76, he served in the agriculture ministry at a time when efforts were made to carry out agrarian reform. In 1977–79, Arias studied at the London School of Economics. He returned to Central America in 1979–80, when he served briefly as an adviser to the Sandinista government. He then spent nine years in Mexico and earned a doctorate in economics from the University of Paris VIII. In 1989, he moved to Panama to direct the PFSA (interview, Salvador Arias, El Salvador, 11 August 1995).

14. The PFSA's initial contacts tended to be with Left-leaning organizations. However, as processes of national-level unity proceeded, representatives of centrist (and occasionally conservative) organizations also attended regional peasant meetings. One of the two Honduran representatives elected to ASOCODE's first coordinating commission, for example, was Víctor Cálix, a leader of the conservative National Peasant Council (ASOCODE 1991b, 31; interview, Víctor

Cálix, CNC, Tegucigalpa, 28 July 1994). The centrist Costa Rican union UPANACIONAL was also involved in ASOCODE from the beginning.

15. *El corte de chaleco* (vest cut) is an untranslatable and eminently Nicaraguan expression that originally referred to a method of executing prisoners and traitors said to have been used by Sandino's forces during the 1927–32 war against the U.S. occupation (and after the marines displayed the heads of captured *Sandinistas*; see Black 1988, 44).

16. Interview, Sinforiano Cáceres, Managua, 4 July 1994.

17. Interview, Rubén Pasos, Managua, 6 July 1994.

18. The Honduran campesino movement, despite its large size and deep historical roots, was severely divided. In 1991–92 negotiations over the agricultural modernization law exacerbated splits between opponents and supporters of the government of Rafael Leonardo Callejas. In an effort to secure peasant backing for the measure, Callejas provided conservative peasant leaders and their organizations' base groups with considerable state resources, including public-sector jobs, vehicles, and promises of land titling and technical assistance. As a result, several large Honduran organizations divided, with one part remaining in the anti-Callejas coalition COCOCH and the other joining the pro-Callejas UNC. In 1994, the two sectors began discussions about the possibilities of reuniting, in part because groups that had supported the agricultural modernization law now sought to amend provisions that had negatively affected their constituencies (interviews, Marcial Reyes Caballero, UNC, Tegucigalpa, 27 July 1994, and Víctor Cálix).

In Panama, key organizations were especially weak after 1989. The Confederation of Agrarian Reform Projects (CONAC) had close links to populist military leader Omar Torrijos and then to the regime of his successor, Manuel Antonio Noriega, who was overthrown in the U.S. invasion. Following the intervention, CONAC was a target of considerable repression (interview, Julio Bermúdez, APEMEP, Panamá, 27 June 1994; Leis 1994, 104–5).

19. The CNA also included COOPEAGRI, a multiple service cooperative in the southern town of Pérez Zeledón, which owns a modern supermarket, coffee processing facilities, and a large agricultural supply house and is probably the most successful peasant-run enterprise in the country. The CNA expanded significantly in 1994 to include several local peasant organizations and a peasant agroforestry group (Voz Campesina 1995).

20. "The Honduran case was very sharp and clear [*tajante*]," CCJYD leader Carlos Hernández recalled. "The campesino organizations didn't expect [the SAP]. When we told them about the impact of adjustment in Costa Rica and what neo-liberalism was going to mean in [the rest of] Central America, they thought it impossible that this could affect the Honduran agrarian reform. . . . This was a very clear position of the peasant leaders we saw in the [PFSA] seminars. They said the agrarian reform was a conquest of the people, that there were laws, that it would never happen, that we were crazy." (For details about the Honduran agrarian reform from the 1960s and the agricultural modernization law of 1992, see Sierra and Ramírez 1994; Thorpe et al., 1995.)

21. Interviews with Central American social scientists, San José and Panama City, June 1994.

22. In April 1990, UNAG resolved that henceforth its union work would no longer be political (in the sense of having party links), but would center exclusively on advocacy for the agricultural sector and the development of the union's enterprises.

23. "If unions [*gremios*] and their leaders work through parties to defend their interests, you get the results you see in Nicaragua. They end up cursing [*puteando*] the commandantes, cursing the parties, and fighting among themselves. We were able to understand that, and fortunately we separated the two things, party and union. . . . The things that have happened to the Nicaraguan *compañeros* have been because of too much intimacy between union and party" (interview, Eulogio Villalta, ADC, San Salvador, 12 July 1994).

24. Ethnic divisions were also more pronounced than elsewhere in the region. Much commercial agriculture was in the hands of Mennonites, who were often not well liked by the English-speaking (and frequently urban) Afro-Belizeans and the Hispanicized Kekchi Maya. The latter groups also felt pressured by the many Guatemalan and Salvadoran refugees who entered Belize in the 1980s and who competed for land and government services. ASOCODE finally attracted Belizean participants after it sent emissaries to identify leaders and cooperative organizations (interview, Julián Avila, BFAC, Panama, 23 June 1994). The Belizean representatives were exclusively from the Hispanic (and Hispanicized Mayan) population. Nonetheless, they were often less comfortable speaking Spanish than English and, especially at first, found it hard to understand many of the technical and political discussions of the other Central Americans (interview, Rodolfo Tzib, CCC-B, Panama, 24 June 1994; Candanedo and Madrigal 1994, 36, 104).

25. Interviews with Guatemalan campesino leaders, Panama, June 1994, and New York, December 1994. All Guatemalans interviewed for this project requested anonymity. In a meeting of ASOCODE's coordinating commission that I observed in June 1994, one of the Guatemalan representatives became irate because his name appeared in a draft of the association's newsletter; if published, this could, he said, have caused him serious problems at home.

26. This mention of biotechnology is one of several factors that suggest that Salvador Arias played a significant role in drafting ASOCODE's 1991 "productive strategy." Arias wrote his doctoral dissertation on the potential dangers of biotechnologies for Central America (Arias 1990). In over forty-five in-depth interviews with Central American campesino activists in 1994 (and many more in Costa Rica in 1988–93), the subject of biotechnology did not come up once, suggesting that this was not a major peasant concern.

27. The "bands" set price ceilings and floors for key basic grains and common tariffs on extraregional imports that were well below the levels already established in the GATT negotiations (see Segovia 1993; Solórzano 1994). Campesino leaders who had attended PFSA seminars participated in the regional meetings of agriculture ministers that led up to the basic grains free-trade agreement. These discussions in 1991 were so heated that the ministers and World Bank representatives "asked for the head" of PFSA Director Arias. He recalled that "the campesinos by now had been studying this for two or three years and they started a confrontation with the ministers and wiped them out [*los anularon*]. The campesinos took apart all their arguments. [The ministers] were unable to respond." CADESCA's director, under pressure from the angered ministers, had to urge the campesino leaders to be more diplomatic in future negotiations (interview, Eduardo Stein). Even though several Central American foreign ministers began to apply pressure for Arias's removal from the PFSA, the representatives from Mexico, Colombia, Venezuela, and Panama on CADESCA's board, as well as the influence of the EU, thwarted their efforts.

28. Individuals and groups at the congress are listed in ASOCODE 1991b. Also present were observers from the CCC-CA, small peasant organizations in Mexico and Cuba, and the International Federation of Agricultural Producers (FIPA), an umbrella group with member organizations in fifty-five countries, headquartered in France; representatives of the diplomatic corps, development agencies, the Catholic Church, and the Nicaraguan and Honduran governments; and CADESCA functionaries.

29. The participating national coalitions were APEMEP (Panama), CNA (Costa Rica), UNAG (Nicaragua), ADC (El Salvador), COCOCH (Honduras), and BFAC and CCC-B (Belize). A range of Guatemalan organizations—including CUC—attended the congress but asked for observer status, since they had not yet founded a national coalition to participate in ASOCODE. CONAMPRO, the coalition that came to represent Guatemala in ASOCODE, was founded shortly after the congress; it was, however, affected by factional disputes and by the withdrawal of its largest constituent organizations, CUC and CONIC (interviews with two CONAMPRO leaders, Panama, June 1994). CUC and CONIC withdrew as a result of disagreements over NGO support for CONAMPRO and the high priority that ASOCODE gave to agricultural rather than political and human rights concerns (Candanedo and Madrigal 1994, 41). Importantly, CADESCA leaders apparently opposed the participation of CUC, because its identification with the armed Left would cause problems with the Guatemalan government (interview, Carlos Hernández).

30. Campos came from a rural community near the central Costa Rican city of Heredia. In a 1990 interview, he recalled that his father had chosen him as the one child out of eight who would attend university. He completed two years at the University of Costa Rica, dropped out to take a position with the Health Ministry in a remote northern zone, and, in the early 1980s, led the formation of the Peasant Union of Guatuso (UCADEGUA), a member of the CCJYD (interview, San José, 1 August 1990). In 1996, Sinforiano Cáceres, a Nicaraguan UNAG leader, succeeded Campos as ASOCODE's general coordinator.

31. ASOCODE's statutes specified that the coordinating commission's decisions must be consensual, rather than by majority vote (1991b, 30). Members described this practice with a mixture of pride (and even wonder) at their capacity for dialogue and frustration at the heated and sometimes inconclusive nature of the discussions.

32. Interviews, Inés Fuentes, COCOCH, Panama, 24 June 1994, and Tegucigalpa, 28 July 1994.

33. Interview, Wilber Zavala, ASOCODE, Managua, 10 August 1995. ASOCODE's statutes provide for the possibility of the associated coalitions paying dues to ASOCODE (1991b, 26) and its 1993 congress approved in principle a dues payment from each national group (Candanedo and Madrigal 1994, 162). Nonetheless, the flow of resources has been exclusively in the other direction, from ASOCODE to the national groups.

34. The creation of a "parallel" typically involves staged elections for a new board of directors. Government agencies or the courts then award the organization's "legal personality" (along with offices, bank accounts, and other resources) to a favored faction, whether or not it is representative of the membership (see Arita 1994; Lombraña 1989; Menjívar et al. 1985; Posas 1985; Thorpe et al. 1995, 131–43; see also note 18).

35. This also occurred in Honduras and El Salvador. ASOCODE leaders wanted to

be able to tell threatening Guatemalan police or immigration officials not only that they had met personally with the president but also that he had stated his approval of their organization. Similarly, connections with ASOCODE (and ASOCODE's ties to European governments) constituted an important form of protection for national groups in the countries where repression of the peasant movement was still common.

36. Interview, Panama, June 1994.
37. Interview, ADC, San Salvador, 18 July 1994.
38. In Honduras, however, the peasant march forced the creation of a new bipartite (peasant organization and public sector) commission to monitor Central Bank credits provided to the state development bank.
39. Interview, Wilbur Zavala.
40. For example, on 23 October 1995, Honduran troops killed three and wounded two in the Department of Yoro when they fired into a crowd of seventy campesinos who refused to leave land claimed by the Ministry of Natural Resources. On 20 February 1996, Guatemalan national police agents violently evicted several hundred campesinos in a CONAMPRO-linked organization who were occupying land in San Lucas Tolimán, Sololá.
41. Another internal ASOCODE document offered a franker assessment, describing the group's reliance on foreign cooperation as one of its "original sins." It went on to say that such funding "is one of the temptations that we will have to face on a daily basis in order to guarantee that ASOCODE has full autonomy and is really at the service of the small and medium-size farmers of the isthmus" (Hernández Cascante 1992, 6). Dependence on foreign funds was a significant preoccupation raised by a number of ASOCODE participants in a recent external evaluation of the organization (Candanedo and Madrigal 1994).
42. ASOCODE sought to guarantee the rotation of cadres in top posts, barring the general coordinator and coordinating commission members from more than two terms in office. Nevertheless, the relative scarcity of skilled organizers in both the national coalitions and the regional farmer and civil society alliances of which ASOCODE is a part (including ICIC, FAC, CICAFOC) suggests that successful leaders can continue their careers well after service to any one organization.
43. ASOCODE's second congress in December 1993 resolved to place the two coordinating commission representatives from each country on the ASOCODE payroll, since they were devoting most of their time to the regional organization. Some national coalitions later objected that this should have been discussed first at the national level, since these individuals also worked for national- and base-level organizations. Some objected as well to a lack of transparency in approving the draft budget, which had not been distributed sufficiently in advance to permit detailed study and discussion (Candanedo and Madrigal 1994, 108).
44. Interviews, Alfonso Goitia, FUNDE, San Salvador, 19 July 1994; Ismael Merlos, COACES, San Salvador, 20 July 1994.
45. Increasingly, though, these concerns are shared and debated. One internal ASOCODE report on a European tour reported that "The ecological issue . . . is one of the problems of most concern to European civil society. Some groups tend to push us towards changes in our cultivation practices that are too drastic. We told them that we were not prepared for this and proposed a more moderate approach toward chemical-free agriculture . . . introducing new practices little by little to achieve a gradual change" (1992b, 4).

References

Acuña Ortega, Víctor Hugo. 1993. Clases Subalternas y Movimientos Sociales en Centroamérica (1870–1930). In *Historia General de Centroamérica. Tomo IV: Las Repúblicas Agroexportadoras (1870–1945)*, edited by V. H. Acuña. Madrid: Sociedad Estatal Quinto Centenario and Facultad Latinoamericana de Ciencias Sociales.

Appadurai, Arjun. 1990. Disjuncture and Difference in the Global Cultural Economy. *Public Culture* 2 (2): 1–24.

Arias Peñate, Salvador. 1989. *Seguridad o Inseguridad Alimentaria: Un Reto para la Región Centroamericana. Perspectivas para el Año 2000.* San Salvador: UCA Editores.

Arias Peñate, Salvador. 1990. Biotecnología: Amenazas y Perspectivas para el Desarrollo de América Central. Ph.D. diss., Departamento Ecuménico de Investigaciones, San José, Costa Rica.

Arias Peñate, Salvador, Juán Jované, and Luis Ng. 1993. Centro América: Obstáculos y Perspectivas del Desarrollo: Un Marco Cuantitativo—MOCECA, Modelo de Coherencia Económica del Istmo Centroamericano. San José, Costa Rica: DEI.

Arita, Carlos. 1994. El Movimiento Campesino: Situación Actual. *Boletín Informativo de Honduras* 68 (July): 1–14.

ASOCODE. 1991a. Estrategia Productiva de los Pequeños y Medianos Productores del Istmo Centroamericano. Mimeo.

ASOCODE. 1991b. Memoria [Primer Congreso ASOCODE] 4, 5 y 6 de diciembre de 1991, Managua. Mimeo.

ASOCODE. 1992a. Informe Congreso Constitutivo. Mimeo.

ASOCODE. 1992b. Informe General. Gira a Europa. Mimeo.

ASOCODE. 1993a. Memoria II Congreso General de ASOCODE. Mimeo.

ASOCODE. 1993b. Plan Trabajo Operativo 1993. Mimeo.

ASOCODE. 1993c. Propuesta para una Estrategia de Cooperación entre ASOCODE y la Cooperación Solidaria con Centroamérica. Síntesis de las Areas Temáticas Priorizadas por ASOCODE. Mimeo. Para Primera Conferencia Regional Campesina sobre Cooperación Solidaria, Panamá, 17–19 March.

ASOCODE. 1994a. Documento Base de Trabajo para el Taller de Seguimiento a la I Conferencia Regional Campesina sobre Cooperación Solidaria. Mimeo.

ASOCODE. 1994b. Memoria del Primer Consejo Regional Campesino Celebrado en Belize (borrador). Mimeo.

ASOCODE. 1994c. Memoria. Taller Regional de Comercialización. Managua, Nicaragua, 09 y 10 de diciembre 1994. Mimeo.

ASOCODE. 1995. Proceso de Planificación ASOCODE Consolidado por Ejes y por Países. Managua, Nicaragua, Enero 04 de 1995. Mimeo.

Biekart, Kees, and Martin Jelsma. 1994. Introduction. In *Peasants Beyond Protest in Central America: Challenges for ASOCODE, Strategies Towards Europe*, edited by K. Biekart and M. Jelsma. Amsterdam: Transnational Institute.

Black, George. 1988. *The Good Neighbor: How the United States Wrote the History of Central America and the Caribbean.* New York: Pantheon.

Bourgois, Philippe. 1989. *Ethnicity at Work: Divided Labor on a Central American Banana Plantation.* Baltimore: Johns Hopkins University Press.

Bustelo, Eduardo S., Andrea Cornia, Richard Jolly, and Frances Stewart. 1987. Hacia un enfoque más amplio en la política de ajuste: ajuste con crecimiento y una dimensión humana. In *Políticas de Ajuste y Grupos más Vulnerables en América Latina: Hacia un Enfoque Alternativo,* edited by Eduardo S. Bustelo. Bogotá: Fondo de las Naciones Unidas para la Infancia.

CADESCA. 1990. *Los Efectos de la Política Macroeconómica en la Agricultura y la Seguridad Alimentaria.* Panama: CADESCA.

Calderón, Vilma de, and Clemente San Sebastian. 1991. Caracterización de los Productores de Granos Básicos en El Salvador. Panamá and Paris: CADESCA, Comisión de la Comunidades Europeas, Gobierno de Francia.

Candanedo, Diana, and Víctor Julio Madrigal. 1994. Informe Final. Evaluación Externa de ASOCODE. Período julio 91-diciembre 93. Mimeo.

CCOD. 1990. *Cooperación externa y desarrollo en Centroamérica. Documentos de la II Consulta Internacional de Cooperación Externa para Centroamérica*, San José: CECADE.

Cohen, Jean, and Andrew Arato. 1992. *Civil Society and Political Theory.* Cambridge, Mass.: MIT Press.

Comisión Centroamericana. 1991. Posición ante el Programmea de Ajuste Estructural, las Relaciones con las ONGs Locales y la Cooperación Internacional Solidaria. Mimeo.

Conferencia Internacional. 1994. Memoria, Conferencia Internacional de Paz y Desarrollo en Centroamérica, PNUD-Plan Especial de Cooperación Económica para Centroamérica, Tegucigalpa, Honduras, 24–25 October 1994.

Consejo Nacional. 1991. La Urgencia del Desarrollo Exige Concertar. Posicion de las Organizaciones de los Pequeños y Medianos Productores del Istmo Centroamericano ante la Cumbre Presidencial Agropecuaria de Centroamérica y Panamá. Mimeo.

Dévé, Frédéric. 1989. Los Productores de Granos Básicos en el Istmo Centroamericano: Ensayo de Síntesis, Logros y Perspectivas Programmea de Seguridad Alimentaria CADESCA/CEE, Guatemala. Mimeo.

Edelman, Marc. 1991. Shifting Legitimacies and Economic Change: The State and Contemporary Costa Rican Peasant Movements. *Peasant Studies* 18 (4): 221–49.

Edelman, Marc. 1994. Three Campesino Activists [interviews with Leoncia Solórzano, Honduras; Wilson Campos, Costa Rica; Sinforiano Cáceres, Nicaragua]. *NACLA Report on the Americas* 28 (3): 30–33.

Escobar, Arturo. 1992. Culture, Practice and Politics: Anthropology and the Study of Social Movements. *Critique of Anthropology* 12 (4): 395–432.

Escobar, Arturo. 1995. *Encountering Development: The Making and Unmaking of the Third World.* Princeton, N.J.: Princeton University Press.

Fallas, Helio. 1993. *Centroamérica: Pobreza y Desarrollo Rural ante la Liberalización Económica.* Costa Rica: IICA.

Feierman, Steven. 1990. *Peasant Intellectuals: Anthropology and History in Tanzania.* Madison: University of Wisconsin Press.

FONDAD. 1993. Campesinos y Ajustes Estructurales. Informe Encuentro Campesino: Pequeños y Medianos Productores de Panamá. Foro sobre la Deuda y el Desarrollo, Panamá.

Garst, Rachel, and Tom Barry. 1990. *Feeding the Crisis: U.S. Food Aid and Farm Policy in Central America.* Lincoln: University of Nebraska Press.

Goitia, Alfonso. 1994. Acceso a la Tierra en Centroamérica. In *Alternativas Campesinas: Modernización en el Agro y Movimiento Campesino en Centroamérica*, edited by K.-D. Tangermann and I. Ríos Valdés. Managua: Latino Editores/CRIES.

González, Vinicio. 1978. La Insurrección Salvadoreña de 1932 y la Gran Huelga

Hondureña de 1954. *Revista Mexicana de Sociología* 60 (2): 563–606.

Gordon-Drabek, Anne. 1987. Development Alternatives: The Challenge for NGO. An Overview of the Issues. *World Development* 15 (autumn supplement): ix–xv.

Hernández Cascante, Jorge Luis, 1992. Para la Evaluación del Congreso Constitutivo de ASOCODE. 6 de enero de 1992. Mimeo.

Hernández Cascante, Jorge Luis. 1994. ASOCODE: Los Retos y Perspectivas del Movimiento Campesino Centroamericano. In *Alternativas Campesinas: Modernización en el Agro y Movimiento Campesino en Centroamérica*, edited by K.-D. Tangermann and I. Ríos Valdés. Managua: Latino Editores/CRIES.

IDB. 1989. *Economic and Social Progress in Latin America 1989 Report*. Washington, D.C.: IDB.

Jelin, Elizabeth. 1990. Citizenship and Identity: Final Reflections. In *Women and Social Change in Latin America*, edited by E. Jelin. London: Zed.

Kearney, Michael. 1991. Borders and Boundaries of State and Self at the End of Empire. *Journal of Historical Sociology* 4 (1): 52–74.

Kruijt, Dirk. 1992. Monopolios de Filantropía: El Caso de las Llamadas 'Organizaciones No-gubernamentales' en América Latina. *Polémica* 16 (January–April): 41–47.

Landsberger, Henry A., and Cynthia N. Hewitt, 1970. Ten Sources of Weakness and Cleavage in Latin American Peasant Movements. In *Agrarian Problems and Peasant Movements in Latin America*, edited by R. Stavenhagen. Garden City, N.Y.: Anchor-Doubleday.

Leis, Raúl. 1994. Panamá: Movimientos Campesinos, Transitismo y Democracia. In *Alternativas Campesinas: Modernización en el Agro y Movimiento Campesino en Centroamérica*, edited by K.-D. Tangermann and I. Ríos Valdés. Managua: Latino Editores/CRIES.

Lofredo, Gino. 1991. Hágase Rico en los 90. *Chasqui* 39:87–91.

Lombraña, Martiniano. 1989. Historia de las Organizaciones Campesinas de Honduras. La Ceiba. Mimeo.

Martínez, Alberto. 1990. *Costa Rica: Política y Regulación de Precios en Ggranos Básicos*. Panamá and Paris: CADESCA, Comisión de la Comunidades Europeas, Gobierno de Francia.

Melucci, Alberto. 1989. *Nomads of the Present: Social Movements and Individual Needs in Contemporary Society*, London: Hutchinson Radius.

Mendiola, Haydée. 1988. Expansión de la Educación Superior Costarricense en los 1970's: Impacto en la Estratificación Social y en el Mercado de Trabajo. *Revista de Ciencias Sociales* 42:81–98.

Menjívar, Rafael, Sui Moy Li Kam, and Virginia Portuguez. 1985. El Movimiento Campesino en Honduras. In *Movimientos Populares en Centroamérica*, edited by D. Camacho and R. Menjívar. San José: EDUCA.

O'Kane, Trish. 1992. Movimiento Campesino: Sembrando Futuro. *Pensamiento Propio* 10 (87): 2–4.

Olofsson, Gunnar. 1988. After the Working Class Movement? An Essay on What's "New" and What's "Social" in the New Social Movements. *Acta Sociologica* 31 (1): 15–34.

Ortega, Emiliano. 1992. Evolution of the Rural Dimension in Latin America and the Caribbean. *CEPAL Review* 47:115–36.

Pelupessy, Wim. 1993. *El Mercado Mundial del Café*. San José: Departamento Ecuménico de Investigaciones.

Pino, Hugo Noé, and Andrew Thorpe, eds. 1992. *Honduras: El Ajuste Estructural y la Reforma Agraria*. Tegucigalpa: Centro de Documentación de Honduras.

Posas, Mario. 1985. El Movimiento Campesino Hondureño: un Panorama General (Siglo XX). In *Historia Política de los Campesinos Latinoamericanos*, Vol. 2, edited by P. González Casanova. Mexico City: Siglo XXI.

Presidentes Centroamericanos. 1991. Plan de Acción para la Agricultura Centroamericana PAC, 15 y 16 de julio. San Salvador, El Salvador. Mimeo.

Presidentes Centroamericanos. 1992. El Compromiso Agropecuario de Panamá, Mimeo.

Román Vega, Isabel. 1994. Costa Rica: Los Campesinos también Quieren Futuro. In *Alternativas Campesinas: Modernización en el Agro y Movimiento Campesino en Centroamérica*, edited by K.-D. Tangermann and I. Ríos Valdés. Managua: Latino Editores/CRIES.

Rosa, Herman. 1993. *AID y las Transformaciones Globales en El Salvador*. Managua: Ediciones CRIES.

Santos, Eduardo A. 1988. La Seguridad Alimentaria Mundial y el Proteccionismo Agrícola. *Comercio Exterior* (Mexico) 38 (7): 635–44.

Segovia, Alexander. 1993. Mercado de Alimentos y Sistema de Banda de Precios en Centroamérica. *Cuadernos de Investigación* (Centro de Investigaciones Tecnológicas y Científicas, Dirección de Investigaciones Económicas y Sociales, El Salvador) 4 (17): 1–20.

Sierra Mejía, Marcio, and Manuel Ramírez Mejía. 1994. El papel del Estado en el Desarrollo del Sector Rural de Honduras hacia el Año 2000. In *¿Estado o mercado? Perspectivas para el Desarrollo Agrícola Centroamericano hacia el Año 2000*, edited by H. Noé Pino, P. Jiménez, and A. Thorpe. Tegucigalpa: POSCAE-UNAH.

Sojo, Carlos. 1992. *La Mano Visible del Mercado: la Asistencia de Estados Unidos al Sector Privado Costarricense en la Década de los Ochenta*. Managua: Ediciones CRIES.

Solórzano, Orlando. 1994. *El Impacto del Sistema Arancelario Centroamericano (SAC) Sobre el Sector Agropecuario: una Aproximación*. San José: IICA.

Stahler-Sholk, Richard. 1990. Ajuste y el Sector Agropecuario en Nicaragua en los 80: Una Evaluación Preliminar. In *Políticas de Ajuste en Nicaragua*, edited by M. Arana et al. Managua: CRIES.

Thorpe, A., H. Noé Pino, P. Jiménez, A. L. Restrepo, D. Suazo, and R. Salgado. 1995. Impacto del Ajuste en el Agro Hondureño. Postgrado Centroamericano en Economía y Planificación del Desarrollo, Tegucigalpa.

Torres, Oscar, and Hernán Alvarado. 1990. *Política Macroeconómica y sus Efectos en la Agricultura y la Seguridad Alimentaria. Caso: Costa Rica*. Panamá City: CADESCA.

Voz Campesina. 1995. La Mesa Nacional Campesina. *Voz Campesina* (Nicaragua) 4 (January–February): 15.

Warman, Arturo. 1988. Los Estudios Campesinos: Veinte Años Después. *Comercio Exterior* (Mexico) 38 (7): 653–58.

9

Conclusion: A Cautiously Optimistic Tale

Jutta Blauert and Simon Zadek

In the sphere of agriculture and rural development in Latin America, mediation for sustainability has become more prevalent in recent years. In the mid-1990s, international institutions and governmental agencies increasingly began to offer another "face" of development, one that envisages policy sensitized to the knowledge and needs of poorer groups in rural communities. In Latin America, as elsewhere, this new face coexists, of course, with the imperatives of structural adjustment policies and related processes. This in itself encourages cynicism and anger in equal measure from those whose experience in rural areas leads them to militate against the local impact of macro-level adjustment processes. Despite this, there are few who offer disengagement as a serious long-term strategy, rather than an as element of the negotiation process itself.

Negotiation using a variety of mediation processes is happening, as institutions at the macro level are beginning to offer something that moves closer to meeting the demands for participation in the debate and policy formation process about sustainable development. It is not only particular individuals in mainstream institutions who are furthering policies that are more acceptable to grassroots interests, although such individuals have, and will continue to have, an important role to play. This is the case particularly with many social, biological, and agricultural scientists who have entered governmental institutions at the managerial level since the late 1980s in Chile, and markedly so over the last five years in Mexico, Bolivia, Colombia, and Central America. After all, whether rural communities, nongovernmental organizations (NGOs), or the World Bank, these institutions and organizations are made up of divergent stakeholder groups, of people with multiple, complex, and interrelated identities, which creates many different opportunities for interinstitutional collaborative action.

In many parts of the world, there are now signs of change in the wider political context aiding this process, such as the energies and initiatives unleashed through Local Agenda 21, with its origins in the Rio Summit.

Provincial and municipal councils, keen on establishing agendas to legitimize their existence and to draw on the technical expertise of other organizations, have seized the opportunity offered by this new sphere of activities. Additional impetus has been provided by the tendency of central government to off-load responsibility on the regional and local governmental tier, although municipal capacity and regional government in Latin America are generally still too weak to develop and initiate their own development plans.[1]

In the end, in this process of concertation[2]—entered into willingly or because of strong external pressure—all actors are now facing a challenge to their methodological capacity and political expertise. They are engaged in a search for planning and evaluation processes that will allow for the mediation of interests surrounding sustainability and enrich the practice of all policymakers, be they international or small producer organizations.

That these institutions are fulfilling new roles at macro and meso levels is due not only to factors current in the policy arena, or to initiatives by enlightened individuals or lobbying campaigns, but also to the search for new methodologies. New ways of facilitating communication, such as Participatory Rural Appraisal (PRA), with its role as translator of local perceptions and demands; indicators of sustainability, with their potential for producing analysis and a more meaningful understanding of changes; and new computer-based technologies have allowed a bridging of spheres of communication. Methodologies that seek to measure the impact of projects for sustainable agriculture and rural development (SARD) are developing rapidly to allow for the appraisal of personal and social dynamics—as well as other less tangible impacts—within grassroots and advisory organizations or even community businesses.

Why Caution?

Real-life optimism must always be accompanied by caution, and the subject matter of this book is no exception. Caution is necessary most of all because of the underlying dialectic of the moment: the emergence of new opportunities to influence policy in favor of SARD, but within a context of an intensification of the practical implications of structural adjustment policies and related processes. The small fingerhold of political power gained by advocates of SARD will need to be consolidated at higher policymaking levels than is the case at present, if it is not to be swept away by events. The alternatives to successful policy influence and effectiveness are unattractive options, such as accelerating or merely continued migration, or the escalation of the power of drug cartels. Effectiveness is needed at all levels, including within local, regional, and national governmental processes, and at the level of international institutions and regulations, including transnational corporations. As a leader of a Central American farmers' association said: "We have forced them to recognize us as a legitimate force. But now, after two years, we've been in four summits and over twenty regional forums. We're seeing that they've made a lot of promises that haven't been kept."[3]

Pressure tactics are now coupled with negotiating skills at ministerial tables and in the boardrooms of multilateral funding agencies and transnational companies. It is necessary for farmers' organizations to forge links with other sectoral groups and with NGOs willing to engage in lobbying work with these farmers' organizations. These forms of transnational organization are increasingly able to forge the development of new markets that can support the poor. However, the resource-poorest farmers, without a competitive commodity to offer, are likely to be left behind such productive international alliance building. The prospect of a concerted action for SARD on their doorstep seems further away. Their only alternative may well be to rely on migrant workers and emigrants' organizations, particularly in the United States, for a more transnational support for policy change.

What is similar among all these rural actors in the 1990s is their general avoidance of, or unease with, party politics, including that of the conventional leftist centralism, which is historically associated with too many failed causes and too much energy lost. In most cases, alliances with other social groups—functional or real—are no longer avoided. They may be with movements (such as environmental or women's organizations), trade unions from other sectors, or the groupings of international advocacy NGOs that make themselves available to support technical and political change (for example, Geyser in France or Farmers' Link in the United Kingdom). The potential of alternative transnationalism for sustainability is already at its clearest in the realm of mediation. Multidisciplinary lobbying[4] for SARD and the use of creative negotiating and mediating spaces that cut across institutional boundaries have finally given a role to groups and individuals at international negotiating tables.

Policies and Actors in Mediation

The participatory paradigm has over the last decade become more attractive to, and more widely adopted by, a range of actors in supporting projects, initiatives, and policies in the interest of sustainable agriculture and development (World Bank 1994; FAO 1994). The take-up of this key SARD principle by such august institutions as the World Bank has been met with a mixed response. A central concern has become that participation could become principally a means for reinforcing a particular—and rather narrow—view of the efficiency of delivery of preconstructed policies under a different "fig leaf." Participation could also, it is argued, form a basis for developing strong, progressive, democratic institutions in the rural arena working for SARD. It is this distinction that lies at the heart of the challenge of contemporary approaches to PRA that Anderson (1996) makes, over what he sees as the more radicalizing approaches to participatory action research (PAR) in the 1970s. The more reformist approach by multilateral donors, confronted by their own contradictions of insisting on structural adjustment policies while including participatory planning as conditions for the funding of many rural development projects, has come under much criticism from this perspective (see Chapter 1 of this book). However, institutional change does not happen in one revolutionary step, and many practitioners in the SARD arena have decided to strengthen their own performance and accountability while continuing to lobby for structural change at national and international levels (see Chapter 5 of this book).

However, participation alone does not guarantee a sufficiently strong position in negotiating arenas or provide skills in mediation. Elsewhere, Escalante (1996) raises a similar set of issues in his exploration of the pressures on participatory development processes in rural parts of Mexico posed by the free-trade agreement between the United States and Mexico (NAFTA). His argument highlights the fact that participatory processes have in the main been taking place with precisely the same groups—namely, peasant communities—that are expected to be affected most negatively through the implementation of NAFTA. NAFTA and related policies, he concludes, constrain those same participatory approaches, since small-scale producers are left with little room for maneuver. Markets that guarantee high farm gate prices are required to validate the high labor costs needed by organic production or the high input costs incurred in chemical farming. Without economic viability in the short and medium term, even guaranteed participation in a governmental or nongovernmental initiative will end in failure, since people will not find it easy to bear the costs of participation.

Macro policies that lead to low farm gate prices or expensive credit and inputs effectively constrain SARD initiatives at the micro level. In Mexico, the impact of NAFTA on the small farming sector has not been buffered even by the cash subsidies of the PROCAMPO program:[5] migration continues to be the key livelihood strategy for the majority of small farmers from the poorer states, requiring new rural financing programs well beyond those currently envisaged even in the social investment funds (Calva 1993; Gómez C. et al. 1993). In this context, participatory practices appear farcical unless broader policies are applied that offer adequate rural income or other incentives sufficiently strong for people to stay in their communities, to farm or process goods, and to risk investing in longer-term environmental regeneration activities or participatory planning and resource management tasks.

As Bebbington's contribution shows, Bolivia offers an illustration of the most recent experience of participation and sustainability having been at least partially integrated within macro policy initiatives (see also Péres 1996). Here, the government attempted through its law of participation to operationalize elements of participatory rural development planning. However, even this relatively progressive move has taken place within the context of structural adjustment programs that make *effective* implementation extremely difficult. Here, the aim was also to strengthen the weak state within the context of further decentralization, by means of a redefinition of the relation between the state and civil society that sought to allow for decision making at the municipal level. In fact, many of the demands and experience by NGOs regarding microregional development planning were integrated into these proposals, presenting ample risks and opportunities at the local and provincial levels—risks of political co-optation and opportunities to apply practical skills. The dual-track policy response to demands for public-sector initiatives conducive to SARD effectively risked dividing the primary sector further than was already the case.

Analysts like Péres (1996) also show that making participation part of this particular style of rural development policy has effectively left the smaller municipal authorities worse off in the context of financing constraints and insufficient technical capacities. These municipalities now have to provide more services but receive relatively less income from central government, given their need to meet the consequent responsibilities for capacity building in administrative and planning skills. Under these circumstances, services contracted out to consultancies and NGOs with long experience in infrastructural service provision could lead to these institutions gaining political power over the weakest local authorities.[6]

Some institutional Mexican experiences (Anderson 1996; Escalante 1996) present at least two related challenges. The first concerns the actual *social* processes underlying the call for and practice of participation that are seen as central to achieving SARD. If these processes are no more than "consultative," argues Anderson, they are likely to achieve very little. The second challenge concerns the specifically *economic* elements of these social processes taking place at the same time as interventions that involve participatory approaches and aim to raise the profile of farmer-to-farmer soil conservation and regeneration practices. Escalante argues that without the right economic policies in place, participation may have little value.

There are clearly actors in place, then, who want to participate in mediation processes and who have acquired ample skills over the last decade. These organizations are also eager to mediate and to support local participation in the SARD arena. However, for participatory processes to engage rural communities in the determination of their own futures, the national and regional contexts need to permit organizational and policy shifts to take place.

Measurement for Mediation

In this context of concern over the political interpretation, practical meaning, and results of participation, the debate about and practice of the measuring of effectiveness against objectives of sustainable development has come of age. The problems associated with relying exclusively on the largely unassessable qualitative strength of participatory processes have led mediators and researchers to focus more directly on the development of new approaches to measurement as a tool for communication. The massive upsurge in work on the development of indicators over recent years has been catalyzed by at least two factors (MacGillivray and Zadek 1996). The first was the fallout from the Rio Earth Summit, which highlighted the need for new approaches to measurement that were capable of capturing the complex interactions of social and environmental phenomena that underpin movements toward or away from a sustainable development path. The Rio fallout inspired institutional support for work on indicators, but it also stressed the matter of participation in the process of defining indicators, as a reinforcement of the governance dimension of the sustainability debate. The second factor underpinning the surge in interest in indicators has been the increased pressure to measure financial and economic effectiveness and efficiency, albeit in the context of the new sustainability agendas.

These two sources of inspiration have proved uncomfortable bedfellows. Agenda 21 has come to symbolize the emancipatory dimension of the measurement process, largely because of its emphasis on the broader definition of sustainability, particularly around the matter of rights (UNDPI 1993). Embodied in this has been the view, "if you want it (specifically nonfinancial factors) to count, count it." The pressure for greater efficiency, in contrast, has often been seen as reflecting an agenda of narrowing the potential for the progressive movement to work on the political dimensions of social change. This perspective can perhaps best be summarized as "if you want it to be funded, prove it!"

The development and use of indicators of sustainable development have therefore been seen by most actors as a critical dimension of success and as playing a central role in influencing policymakers. This is why the focus on more effective indicators as a support to mediation between actors has been recognized and explored in two of the chapters in this book.

At one end of the spectrum are those processes of measurement precipitated mainly as a means of assessing the effectiveness of a particular project or initiative. The chapters do not address the economic modeling of farm households to ascertain the financial effectiveness of sustainable agriculture practices. Economists might argue, as do Ruben and Heerink (1996), that such practices need to be demonstrably economically viable if rural financing institutions and policymakers are to be convinced of their relevance. Indeed, traditional economic analysis *can* point out appropriate paths for policy and practice in a language that is familiar to policymakers, as long as it allows for a wide set of environmental and social indicators of effectiveness.

The experience within development projects is a critical dimension of what needs to be understood. Key issues surrounding this deceptively simple sounding aim are illustrated clearly by the Ecuadorean case analyzed by Does and Arce (1995). From a perspective of social anthropology, the case focuses on the means by which direct, oral narrative can be legitimized within the evaluation process of a project and, through this, guide us to the qualitative end of the indicator spectrum. By seeking out such different operational narratives in a project, outside observers can make clear the different positions adopted among the key actors and thereby assist in ordering and mediating between diverse interests and options, such as in the selection of technology. In assessing the role of the narrative in the process of policy formation, they confirm that discontinuities between formulation and implementation within projects does not imply the lack of connection between the activities, but rather that a different type of connection exists—that of contested knowledge and interests. Mediation, therefore, happens not only from communities or small organizations at the micro level "upward," but also *within* communities and projects. This heterogeneity needs to be acknowledged and worked through, even if it takes time. In this context, they urge a conscious investigation of apparent discontinuities caused by the involvement of outsiders and the changing internal dynamics of SARD projects. These investigations, it is argued, offer a means of mediation between, for example, different factions within rural communities and between resource-poor communities and policymakers at the state level and in international agencies.

Irene Guijt's chapter takes this form of qualitative analysis forward in describing a number of direct interventions in project assessments with a view to strengthening the quality of the often formalistic indicators produced and reported by farmers' organizations or technical projects in watershed management. Whereas authors such as Does and Arce highlight the value of uncovering the meaning of multiple narratives, Guijt stresses the key role of self- or participatory assessment, both as a means of increasing the quality of the evaluation process per se, and as a means of maximizing its application. Guijt concludes that sound evaluation of different areas of sustainability allows the evolution of indicators that can be effective in validating SARD approaches at the policy level and in improving on current knowledge. More importantly, we learn that the transmission mechanism for this enhanced policy effectiveness is not so much the new indicators themselves as the rural institutions (farmers' organizations, local NGOs, or municipal and community authorities) that may be strengthened by this process of indicator development, a subject returned to later.

Methodologies from indicator development as well as anthropological approaches are oriented toward the assessment of reasonably well-bounded sets of events "on the ground"—that is, within rural communities and related specifically to agricultural production. Yet, although implications for policy are offered, and the Brazilian and Central American case studies are presented with considerable self-conscious reflection on the policy effectiveness of their own approaches (as opposed to that of farmers and rural communities), no offer is made for methodologies explicitly designed to assess policy effectiveness. Marion Ritchey-Vance's chapter on the work of the Inter-American Foundation (IAF) in developing the Grassroots Development Framework—or the Cone—does precisely this. The Cone arose from a concern among IAF staff that traditional approaches to evaluation did not allow the organizations' and beneficiaries' experience to be considered in relation to regional or national contexts or individual program strategy. The traditional approaches failed to identify those project gains that were intangible but frequently key to the sustainability over time of a development process. The Cone offers a means for accounting for these intangible and broader effects. It does so by providing a six-element taxonomy of information about effectiveness, covering the tangible and intangible dimensions of direct benefits, organizational effects, and broader societal effects.

Current frameworks of measurement, evaluation, and monitoring are constantly being reassessed, and new ones are evolving. Indicators in the context of the Cone, as with the other methods considered earlier, are but one element contributing to such frameworks. Indicators are not new. What is potentially new is what they are designed to report on, how they are designed, by whom they are used, and to what purpose. The IAF-assisted experience highlights the need to *limit and order* the number of indicators used so as to prevent the emergence of an unmanageable mass that cannot be applied or compared. The teams' discovery of the usefulness of a thematic hierarchy for ordering indicators confirms work elsewhere (Doyal and Gough 1991; Max-Neef 1991).

Indicators are varied and need to be simple and translatable. For instance, the farmer-to-farmer approach to sustainable agriculture and natural resource management is recognized to be both solid in local impact and notoriously difficult to verify for its intangible and invisible impacts.[7] Yet skeptical researchers, agronomists, and commercial farmers still question the potential for scaling up the farmer-to-farmer approach and low-external-input agriculture to a commercial or regional level.[8] To address these queries, work in Latin America is increasingly pointing to the need to include external stakeholders in the definition of indicators from the start of the process in order to appraise the potential sphere of influence and take-up. Setting local views alongside the criteria of external stakeholders—to which indicator and operational narrative work contributes—can aid the process of mediation between conflicting interests by providing an early indicator of conflictive or overlapping expectations on SARD projects (Blauert and Quintanar 1997; Does and Arce 1995). Involving external stakeholders in monitoring and evaluation processes of innovative practices, it is argued, allows a stronger influence over the appraisal of impact over policy in the longer term. Ritchey-Vance argues that funders today aim primarily to downsize their portfolios and thus stress efficiency as a criterion for rating and culling grant recipients. Their interpretation of efficiency frequently relates more to compliance or skill in administrative management than to effectiveness in fulfilling the needs of beneficiaries, building social capital, or achieving concrete change. Without encouraging NGOs to measure their broader effectiveness, the pressure from donors will gradually push them toward a more narrow focus. In allowing this route to be taken, the policy orientation of NGOs and grassroots organizations will degenerate, and "we may kill the goose that is laying the golden eggs" (Ritchey-Vance, Chapter 3 of this book).

A critical element of the struggle to assert indicators of sustainable development is taking place at the national and international levels. Without challenging measures of effectiveness (and associated policies) of entire national and regional economies and their environmental bases, the micro and meso battles fought by rural communities are likely to be lost through the overwhelming influence of macroeconomic effects (MacGillivray and Zadek 1996; Zadek and MacGillivray 1996). From this perspective, it is stressed that SARD is not an isolated phenomenon, and it is equally important to develop and make explicit measures of urban development—or indeed, measures of political freedom—within the policy debate in support of SARD.

The development and application of new indicators should not, however, necessarily be seen as unambiguously advantageous to the efforts to achieve effective SARD. The chapters in this book describe mostly "progressive" initiatives that seek—or have managed—to empower rural actors both by strengthening their capacity for self-reflection and by enhancing their ability to articulate their needs and experiences. Simultaneously, however, the drive for new indicators is coming from institutions focused on narrowing the scope of rural development processes. Indicators can, then, be a double-edged sword that can be turned against the cause of SARD and undermine its resonance in policy debate and practice.

Appropriate indicators are in this sense an absolutely necessary weapon in the struggle to legitimize SARD and to strengthen its profile at the policy level. They are both a tool for redefining the voice of SARD and a space where the underlying struggles of rural communities and their associates may be heard. They offer enormous scope to strengthen mediation between different interests and associated discourses and to bring pressure to bear where necessary on the public and in the professional spheres. Finally, the process of developing and using indicators can offer a means to strengthen progressive institutions, a subject to which we return in a later section of this chapter.

Markets as Mediation

It goes without saying that the economic conditions facing rural populations and their organizations are a critical factor in their ability to achieve sustainable development processes and livelihoods. The chapters in this book do not deal merely with the attitudes or approaches of agricultural research institutes and extension services, or of development agencies more generally; they concentrate also on the terms on which agricultural produce enters the market after production by smallholder farmers. The market is identifiable in Latin America as both the unavoidable source of the crisis emerging from unsustainable practices and a route to addressing this crisis (Linck 1993). In many of the chapters, this context is implicit, particularly in those that focus on measurement issues and political processes. Several of the chapters, however, address the market directly and, in particular, the strategies followed to engage in the market on terms that support the principles and practice of SARD. Of significance throughout the chapters are the roles of mediators in supporting the negotiations of peasant and other rural organizations with the state and with business.

The experiences of building peasant forestry enterprises in the Yucatán Peninsula in Mexico related by Gerardo Alatorre and Eckart Boege provide a starting point for an investigation of this phenomenon. At the heart of the joint Mexico- and Germany-supported initiative that they describe, the Plan Piloto Forestal (PPF), was a view that the destruction of the rain forest could be stopped only by the owners—that is, the peasant communities—participating in the logging, managing, and processing of their resource. At the same time, the authors argue that the NGOs that had traditionally engaged with these communities had little or no interest in engaging with state or commercial institutions at the outset, and they had no experience in either conceiving or implementing social enterprises. Furthermore, the existing institutional structures through which change needed to take place were themselves in a poor state, with heavy bureaucracies, corruption in peasant organizations, and an overwhelming inertia that made progress virtually impossible.

What broke this vicious circle in the early 1980s, argue Alatorre and Boege, was the combination of three developments in the public and professional sectors. First was the emergence of a new generation of agricultural and forestry technicians, who were prepared to create and defend a new professional engagement. Second, and critical to securing their role, was the fact that they were able to operate with a high level of autonomy by virtue of external funding through the German-funded PPF initiative. Finally, their success was also dependent on the effective economic value given to the forest, aided particularly by an increase in cedar and mahogany prices negotiated with the timber companies.

The outcome of this process to date has been a successful development of social forestry enterprises, which has strengthened the capacity and resolve of peasant organizations and individual farmers to maintain the forests in the face of considerable economic pressures and needs. This experience, the authors argue, is now sufficiently well developed to offer practical insights for future forestry policy in Mexico and Latin America. The institutions that emerged through the evolution of regional forestry enterprises were important in the process. This was particularly the case for NGOs, which, the authors argue, have been "cultural translators" or bridges between the peasantry, NGOs, and research, governmental, and aid organizations. At the same time, they have brought closer together in their work both productive and environmental issues, as well as the practical and theoretical domains, within an overall response to both local and global concerns.

This exploration of the role of social enterprise is also relevant to the discussion of the emergence of the fair-trade market for coffee, as demonstrated by Pauline Tiffen and Simon Zadek in Chapter 6. Once again, globalization of the coffee market is seen as lying at the heart of the problem, yet it is within these globalized markets that the response from peasant organizations and their associates has been addressed. The role of a number of European fair-trade organizations is important today in developing fair-trade criteria and quality assurance processes, consumer awareness, and the practice of fair trading itself. At the heart of this commercial relationship and practice, the authors assert, are the primary producers, and not the product, with local skills, natural resources, and context being key to the commercial and developmental success. The alternative trader, it is argued, seeks to differentiate the value of inputs by the different parties in the trading chain, with the aim of maximizing the gains from trade that accrue to the Southern supplier (Zadek and Tiffen 1996).

Policy in this context has many dimensions. The authors show that formal government policy may be the least important factor in determining the success or failure of social enterprises, since government has little direct role in the market, now that many of the commodity-purchasing and input-producing public-sector companies have been privatized. Policy in particular concerns the degree to which fair trade influences mainstream corporate practice by providing a counterdemonstration in the marketplace to normal commercial practice. The evidence offered in Tiffen and Zadek's chapter is limited to some specific export commodities from smallholdings (less perishable agricultural produce). But it offers a window to the potential scope for change in corporate behavior, a type of social actor now under pressure to respond to the fair-trade movement through resort to opening itself to ethical performance criteria. As to the role of the alternative traders in influencing this direction, the authors conclude that they have and will continue to have an important role to play as the "compass"—the "true north" for the movement as it expands.

Tiffen and Zadek's chapter focuses on the role of Northern alternative traders in influencing commercial operations such as Nestlé directly through being *in* the market. Alistair Smith from Farmers' Link in the United Kingdom focuses on mediation strategies for influencing not only business directly but also the European statutory framework through which the global banana market functions. He provides the example of one small organization based in the United Kingdom, Banana Link, in assessing the possible roles of Northern organizations in the mediation for SARD and fair trade.

Banana Link has been key to the creation of a complex process of coordinating an information network and facilitating the building of an international alliance around the objective of sustainability in the banana sector. By presenting a virtual diary of events from 1993 to the present, Smith highlights above all the extraordinary complexity of any process of intervention or mediation in an international market with many institutional players with diverse interests. Written in a consciously personalized style, the author makes clear that mediation for him is a process of persuasion, not in any sense a role involving neutral arbitration. This description most explicitly mirrors the challenge posed by Anderson (1996) as to the radicalizing role of the mediator in encouraging social mobilization. With regard to policy effectiveness, Smith sees the events he describes as having generated the necessary knowledge, institutional development processes, and public interest in Europe for serious negotiations to begin. He concludes that together with certain economic actors from the producing countries, Banana Link is now in a position to start exerting concerted pressure on Europe's only remaining major private-sector banana trading company to cooperate on Banana Link's terms at both economic and political levels.

Smith's vision, however, extends beyond even the large-scale challenges and potential gains from influencing the terms of trade of the banana sector.

He sees the "banana war" as a demonstration of what could be achieved elsewhere. The results of mediating sustainability in the banana sector emphasize the importance of adopting concrete approaches. Without such real-life examples, debate that advocates a new international economic order is likely to remain abstract theory or speculation. But where concerted attempts are made to change the terms of trade, it is possible for the producer and lobbying organizations to better define what actually might be meant by sustainable trade.

The experiences related in these chapters throw a very different light on the role of mediators in the process of translating practice in pursuit of SARD into policy. Policy is clearly acknowledged to extend beyond the traditionally dominant realm of state policy to include the ways in which markets for agricultural produce function at local, national, and international levels. In the cases of the markets for coffee and bananas, international coalitions of producer organizations, traders, and campaigners have been formed to influence the terms of trade on which these products are produced and sold.

In both these cases, moves to influence the state have been seen as instrumental in achieving leverage over the behavior of transnationals themselves, a leverage that governments have in many ways forfeited or lost. These initiatives have taken as their starting point the need to strengthen the capacity of progressive producers and traders to be more effective *within* international markets. Moreover, in both cases, they have taken a total-system approach to their strategy of simultaneously developing consumer awareness of the inequities of dominant trading systems and of the existence of practical alternative approaches that can support the principles and practice of SARD (Barratt Brown 1993).

Organizing for Mediation

Central to the experiences described throughout this book is a recognition of the need for new approaches to organization if effective mediation is to be achieved. The tools or methodologies presented involve either new indicators, and the exploration of narrative dialogue, or the construction of unconventional economic strategies. To be effective, strategies both outside and within the market require a radical reorientation of the terms of engagement with both new and old actors from the state and business. In each and every case reported in the chapters, there has been an evolution in institutional capacity to incorporate those technical, commercial, and negotiating skills required to formulate and implement new strategies to achieve SARD.

In the case of social forestry practices in Mexico, for instance, NGOs have joined forces with technicians and international donors in developing the capacity of peasant organizations to exploit natural resources in a sustainable manner. These actors have used economic as well as political leverage to develop the institutional capacity and will of peasant organizations to acquire greater control over local resources to both achieve economic gain and establish viable stewardship over environmental wealth.

These emerging institutional forms come at a time of increasing skepticism about, and even cynicism toward, the legitimacy and effectiveness of civil institutions, as the critique of NGOs by both funders and grassroots organization highlights (see Chapter 2 of this book; Niekerk 1994; Toranzo 1995). However, these new institutional forms also arise—at least since the outspoken and constructively critical linkages made with development NGOs by the Zapatista insurgents in Chiapas, Mexico, after January 1994—within a context of increasing self-confidence of civil organizations as the state and international funders look to NGOs for professional expertise. Yet the by-now familiar problems associated with rapid institutional growth have not been adequately addressed, either conceptually or in practical terms: careerism linked to professionalization, a deterioration in the democratic process associated particularly with scale, and an increasing incidence of corruption or at least malpractice as resource levels rise. Compared with the 1980s, when the modern structure of international nongovernmentalism passed through its formative stage, the 1990s have seen a growing disaffection and concern over the sector's future (Edwards and Hulme 1995; Sogge 1996).

Anthony Bebbington's chapter on the changing roles of NGOs as mediators and even direct policy actors and researchers is relevant to Latin America beyond the Andean region on which he bases his analysis. Since the early 1990s, similar experiences have been lived in most other Latin American countries. NGOs have to rise to new challenges to their skills and professionalism, and they find themselves suddenly in a stronger position in their capacity as service deliverers, for example, to the formerly marginalized municipal and provincial governments. Yet generalizations for the region as a whole are inappropriate, and it would be dangerous to take those proposed too seriously. In Bolivia, where the state has traditionally been weaker than in countries such as Chile or Mexico and where, conversely, NGOs have a longer experience of professionalized service provision, the opportunities for mediation and institutional change for SARD are led by external actors (funders) and NGOs rather than by pervasive central governmental policies that influence almost all action at the rural level. Péres (1996), for instance, describes the interinstitutional councils for microregional planning that have been brought into existence by European development agencies and their NGO counterparts in Bolivia. Contributing to policy formulation, implementation, and monitoring, these institutions—or spaces for concertation—are now made up of governmental representatives, NGOs, funders, rural organizations, and local authorities. The mediation role of NGOs can thus be set within the ambit of a broader collaboration, and, it could be argued, in the case of the microregional development groups, funders have helped create institutions that allow and indeed enable mediation between local and national concerns.

The new area to be watched in relation to SARD is the municipal level and the possible triangular relationship between NGOs and research institutions, municipal authorities, and communities or social enterprises (see Thrupp 1996, 14ff.). The recent exploration of possible collaboration along these lines—which developed between AS-PTA, local farmer organizations, and municipal authorities in Brazil (see Chapter 4)—is echoed by the planning groups in Bolivia (Grupo DRU), interinstitutional collaborations in central Chile, and initiatives in Mexico for regional sustainable rural development plans (PRODERs).

Although peasant and more generally community-based organizations have entered the limelight as the real basis of a civil movement, concern about their democratic credentials and potential for leadership is now beginning to emerge, just as the legitimacy of many NGOs is being questioned (see Chapter 8). However, these well-documented problems within many civil institutions[9] in no way reduce the imperative for creative, energetic entities that are both responsive to grassroots needs and capable of engaging with a range of existing actors who are central in setting the relevant policy frameworks. Combined with the increasing pressure on their intended recipients and beneficiaries as a result of the overarching process of globalization, it is precisely the shortcomings of many institutions emerging from the 1980s that have made more urgent the need for the development of new institutional frameworks through which new social movements can articulate their perspectives.

In this paradoxical and conflictive context, the intentions of and the roles adopted by the mediator become absolutely critical in determining the nature of the institutions emerging from any process. Usually, it is only the "higher" levels of participation (interactive participation, engaged participation, and participation for social mobilization) that support the emergence of strong and potentially effective democratic organizations (Anderson 1996). In addition, macroeconomic policies go a long way in predetermining the quality and effectiveness of participatory processes, since they are likely to fail if they seek to build grassroots institutions and programs based on economic projects doomed to failure by virtue of their macro context (Escalante 1996). The fortunes of institutional development processes are therefore partly contingent on the broader policy framework and yet are simultaneously the basis for their transformation.

Recognizing the different institutional and personal contexts of the role of the mediator, Alistair Smith in Chapter 5 presents us with two key definitions of different actions in mediation:

- *Intermediation* encompasses a range of deliberate, or conscious, unilateral and multilateral interventions motivated by common educational, economic, and political objectives. *Educational mediation* can be either popular or academic; *economic mediation* can either be analytical or can involve becoming an economic actor in the sphere in which one intervenes; and *political mediation* can be based around critique or proposal, or both

- *Intramediation* is more about the separate but interlocking process of consciously gaining an understanding of the different actors and their interests within the evolving network and alliance. This aspect of the still ill-defined concept of mediation seems to be a prerequisite for mediation in solidarity

A more practical definition of "mediation," Smith argues, might include the facilitation of exchanges between people and of information, the creation and nurturing of forums for informed debate, the construction of politico-economic alliances, the involvement of farmers and farm workers in research or in challenging existing institutions, or any combination of these activities.

Smith offers then, a range of different possible actions: interventions as well as accompaniment and joint learning. Effectively, he opens the way for different actors to have their roles in mediation acknowledged, thus inviting institutions and individuals to take responsibility for their actions in this mediation.

These modes and degrees of mediation and participation also have implications for organizational forms and procedures. Issues of organizational development are reflected in the measurement processes reported in a number of chapters. Guijt's description of the impact of indicator development processes led by farmers or undertaken with their advisers focuses on the growth in the reflective capacity of rural institutions and highlights the limitations of such capacity. The imperative for learning within institutions refers to the evolution of both method and institutional form. She argues that monitoring and evaluation with all stakeholders can offer vital opportunities for motivation and the modification of existing efforts in sustainable agriculture (see also Blauert and Quintanar 1997).

Ritchey-Vance's description of the Cone and its use is even more conscious of these two dimensions of institutional effect. Drawing on Putnam's formative work on social capital, Ritchey-Vance supports his conclusion that "[t]he historical record strongly suggests that the successful communities became rich because they were civic, not the other way around."[10] Recent work by the World Bank seems to confirm this view that social capital counts in generating sustainable livelihoods. Returning to the attack on short-termism in funding strategies, she concludes that conventional cost-benefit measures may reasonably reflect the nuts and bolts of development projects but effectively ignore the less visible, less marketable efforts directed at building human capacity and social capital.

Ritchey-Vance's chapter not only highlights the importance of institutional development but also offers a methodology for identifying and thereby validating institutional effects arising from development processes. For example, the second level of the Cone incorporates an assessment of effects on organizational strengthening—whether as NGOs or community institutions. In each case, the application of the Cone in assessing progress revealed significant organizational development effects arising from projects focused principally on other aspects of SARD. But more importantly, the Cone method allows both a systematic analysis and a presentation of process and of the impacts of an organization's activities, where changes occurring at the micro (technical and personal) level give rise to indicators as much as do those at the meso (organizational, communitywide) level. Institutional impacts and required changes can thus be monitored according to the definition of sustainability that each set of program actors has.

The IAF is itself, of course, an actor or institution through which key policies can be influenced. Yet like many other institutions in that powerful position, "despite demonstrated success in crucial areas . . . the lessons learned by IAF have not translated into support for foreign assistance, even of the self-help variety" (Ritchey-Vance, Chapter 3 this book). It is perhaps a sign of the achievement of current analyses of policy impact that the time is now ripe for further research into the degree and means by which agencies such as the IAF have been able to use evaluative and planning innovations such as the Cone to understand their own impact on policy.

Alistair Smith's chapter on bananas, discussed earlier, offers further support for the view that institutional development lies at the heart of building effective mediation processes for supporting SARD. In reflecting on the initial stages of involvement of the U.K.-based organization Farmers' Link (FL), Smith argues that the Caribbean banana producer confederation WINFA (Windward Islands Farmers' Association) realized that it would need to build up its own sources of information and modes of communication if small-scale banana farmers were to have a voice in Europe that was independent of their own governmental lobby and the Banana Growers' Associations that the governments controlled. FL responded to a direct request in St. Lucia for it to play a specifically defined informational, logistical, and networking role in support of WINFA's attempts to build its own institutional capacity.

Like Edelman's, Smith's chapter highlights the evolutionary nature of institutional development processes. From this initial informational contribution, FL deepened its involvement to include a wide range of lobbying, negotiation, facilitation, and educational roles. At each stage of development, Smith reflects, the mediating organization had to sharpen its listening skills and initiate responses to what it thought it heard was needed rather than developing an agenda itself, however well-meaning this might have been.[11] Moreover, the understanding of the scope of institutional development is extended considerably through the experience described in Chapter 5. It includes not only the strengthening of progressive civil institutions but also the transformation of institutions in the state and business sectors. For example, Smith points to the International Network for the Improvement of Bananas and Plantains as being unique within the network of multilateral international research centers (CGIARs) for its relative openness to proposed policy changes and dialogue, thereby offering an opportunity for dialogue with a view to influencing the broader network of international research institutes. As in the case of fair trading in coffee and other commodities, it has been as much the lobbying as the trading skills gained by actors on both sides of the Atlantic that have led to successful policy changes, as the changes in trading regimes (between ACP-EU countries) has proved.

Marc Edelman, in the one chapter in this book that focuses exclusively on institutional and leadership development processes, explores the development of the Association of Central American Peasant Organizations for Cooperation and Development (ASOCODE). In doing so, he offers rich insights into many of the challenges in building a transnational peasant movement capable of engaging effectively with policymakers. Edelman points to the organizational need for peasant leaders who are able to deal on equal terms with their policy counterparts. He quotes one analyst explaining the need to challenge the modes of expression of his peers in a peasant farmer organization:

> I can explain globalization in the simplest way and you will understand me. But if you don't handle the terminology used by the politicians with whom you're negotiating, even if you know about globalization, you're not going to understand them, because they aren't going to use your categories. You have to use their words. When you're negotiating, you can't ask ministers to negotiate at your level. You have to raise the level. And now it's easy to find campesino leaders in Central America who can speak about macroeconomics, about economic adjustment. (Edelman, Chapter 8 of this book)

He points out that beyond the need to reform the modes of expression lies a host of difficult problems. Central among them is the question of how best to manage the professionalization of the peasant leadership, as he sees it, and, in particular, the disjuncture in experience between these leaders and their constituencies that arises through the former's transformed lifestyle, increased access to resources, and international exposure. Related to this is the likely trajectory of the institutions that emerge to support such leadership, in particular ASOCODE itself. At what point, he asks, does its commitment to maintaining a dialogue with policymakers lead it to be interested in controlling the more vociferous grassroots organizations that it purports to represent?

Edelman does not seek any final conclusion to what he sees as a critical set of issues facing a new generation of peasant struggles. As well as pointing out some of the difficulties facing an institution like ASOCODE, he celebrates many of its institutional innovations, including its rejection of any political party affiliation and its willingness to engage with other social actors—including environmentalist, women's, and indigenous groups. Ultimately, he concludes, an organization like ASOCODE—designed to strengthen democratic representation of peasant communities in national and international policymaking forums—may fall foul of exactly the same set of operational and political problems that have beset NGOs in recent years. In that case, he concludes, ASOCODE is likely to find its own legitimacy challenged by a new generation of peasant leaders and organizations, who in turn will seek to reassert democratic control over the mediation process.

The call for institutional development can be likened to the call for nongovernmentalism. There is a sense of the gap that is being filled, but by a category of phenomena that is too broad to achieve operational meaning. Yet the experiences of mediation described in this book do point toward the central importance of new forms of institutional development to underpin effective mediation processes. Disembodied indicators of sustainability alone have neither a clear purpose nor point to a traversable path. Required are institutions that are capable of building indicators on the firm base of participatory processes with rural communities, and that are equally capable of projecting them in policy forums with professionalism. Similarly, the vision of social enterprise or fair trade will not move beyond principles without institutions that are able to cope with the almost schizophrenic condition arising from the combining of commercial venturing with both social and environmental concerns and an underpinning democratic process.

Finally, institutional development is also required at the public-sector level, if some role is still envisaged for the state as both policymaker and even mediator. As the example of the technicians in the southern Mexican case described by Alatorre and Boege showed, and as Escalante (1996) points out for agricultural ministries and financing institutions, there is space, indeed a need, for enlightened policies and public-sector institutions to cater to the practical requirements of participatory and interinstitutional work for SARD.

Seeking Room for Mediation

The experiences described in this book do not, therefore, provide off-the-shelf models for application to SARD initiatives throughout the region. The

lessons of institutional development can be useful only if their specific rec-
ommendation domains are clearly understood (Ruben and Heerink 1996).
Despite this caution, a number of lessons with considerable potential resonance can be elicited from the analyses provided in this book.

Foremost is that the traditional categorization of activities associated with
particular forms of institutions no longer effectively serves the cause of
SARD. For example, to be a campaigning organization can no longer mean
to maintain a homogeneous confrontational style; to be a democratic organization cannot allow one to evade the need for professionals; and to be a
nonprofit, nongovernmental organization does not mean that you cannot
engage in commerce, both to further strategic interests and to secure institutional viability. At a broad level, commitment to SARD in Latin America no
longer means that other dimensions of the broader struggle for sustainable
development in many parts of the world are not an integral part of the
agenda.

These sorts of challenges to the traditional differentiation of roles within
civil society imply that not simply different but also diverse institutional cultures are required *simultaneously* in any one organization, network, or
alliance. The need to create effective interfaces between at least some technicians, World Bank officials, and grassroots activists, for example, speaks to
complex institutional formations that contain and can work with multiple
languages and their associated worldviews, interests, and pressures (Zadek
1995). Furthermore, these institutional formations will themselves be
required to evolve rapidly in the face of shifting circumstances. Gone are the
days when it was possible to create a body to negotiate repeatedly around
issues with organizations that were qualitatively relatively static over time.

This complex, dynamic, and potentially chaotic set of needs places enormous demands on leadership. It highlights a range of challenges concerning
the identification and molding of leadership and possibly the duration of its
"life cycle" prior to becoming subject to legitimate challenge and replace-
ment as its own self-identification reforms away from its base constituency.
Here, perhaps more than anywhere else, many of the chapters do not provide the depth of description and analysis required to understand the revolution in leadership and personal evolution that is happening or is still
required (Edelman's chapter is an exception to this). For Latin America, as
elsewhere, much research is currently being undertaken about this subject,
but as yet, little has been published in which the subject of analysis is a local
or international institution rather than a farmers' or grassroots organization
or a development program.[12]

The organizational forms that will dominate the institutional landscape in
the SARD arena in probably less than a decade from now are likely to be
radically different from the major organizations operating today (such as
NGOs, commercial institutions, public-sector agencies, and producer

organizations). From today's perspective, tomorrow's institutions will look very fragmented and possibly chaotic as the traditional patterns of hierarchy, bureaucracy, formal legal boundaries, and the relative homogeneity of colleagues and partners and their work areas increasingly break down. These changes are likely to take place within the context of a further intensification of globalization and very different processes for distributing what will probably be a far smaller quantity of grant funds to support SARD. Many of today's mediators are likely to play a critical role in the formative development of these new institutions, as the chapters in this book suggest they are already. Indeed, many of today's mediators may prove to be key actors in the leadership of tomorrow's institutions that are to contribute to such a change.

The arguments of and contributions to this book do not aim to be prescriptive of how SARD should be practiced or achieved across the region, or even for one particular area or people. Nor does this book only tell stories of successful local resource management or agricultural production strategies, although there are as many positive experiences and strategies to be described as negative ones.[13] There are many suggestions, however, that have evolved based on many years of personal and secondhand experience, as to how mediation of differing visions of sustainability can be achieved. Above all, the book argues and the contributions show that neither governments, development agencies, farmers' organizations, NGOs—as policymakers and operational agents—nor community forestry enterprises can avoid entering the mediation arena. Clearly, mediation does not do away, for example, with the need to offer technical support for migrant populations arriving from degraded areas to lowland or periurban agroecosystems about which the farmers hold often inappropriate knowledge. There are, however, methods, forms of organization, and even financial resources available that allow improved communication both horizontally (between peer groups) and vertically (between micro and macro levels), as well as a scaling up of relevant experiences that can provide both expertise in technical solutions for SARD and skills in political lobbying, organizational development, and mediation itself.

In the final analysis, the use made of the lessons learned from praxis, or from the innovative side of policy, depends on new forms of economic relations being evolved; on transparent, self-critical appraisal of work being undertaken; and on communication between different actors being enhanced. Although each chapter in this collection offers implications for policy, and some reflect self-consciously on the policy effectiveness of their own approach (as opposed to that of farmers and rural communities), none yet offers methodologies explicitly designed to assess policy effectiveness. There is a space wide open, then, for systematically learning about changes necessary within farmers' organizations, NGOs, and local government procedures if policy is to be effective in supporting SARD objectives. Also, the effectiveness of SARD policies developed, executed, and promoted by farmers' organizations and NGOs turned policymakers would usefully be appraised. Intramediation may be a key step in moving toward our objective: that rural people should be able to define and responsibly manage their own livelihood system while contributing to socially just and environmentally sustainable resource use.

Notes

1. Again, countries such as Chile and Colombia indicate a more speedy uptake of this potential local mediation and empowerment process.
2. Concertation, from the Spanish *concertación*, is the commonly used expression for political negotiation and consensus seeking between different actors. Originally, this word was used only by public-sector institutions, aided by the language of multilateral funding agencies, in seeking the involvement of dissident rural and urban social actors in the proposed policy process. Today, even some NGO sectors accept this word as an expression of the endeavor to find a common purpose and dialogue between different actors in the interest of rural development.
3. Leader of the Central American farmers' organization ASOCODE, cited in Chapter 8 this book.
4. Thanks to Pauline Tiffen for this new term.
5. The Mexican governmental PROCAMPO subsidy program pays smallholders US$55 per hectare of land planted to maize.
6. For an experience similar to the uneven one of the Municipal Solidarity Funds (*Fondos Municipales de Solidaridad*) in southern Mexico, see Fox and Aranda 1996.
7. For a detailed description and critical analysis of these kind of programs in Central America and Africa, see Gubbels 1994; Holt-Giménez 1995.
8. For a much required economic analysis of potential impacts of sustainable agriculture in Central America, see Ruben and Heerink 1996; otherwise, skepticism is echoed across Mexico, Central America, and Andean countries by governmental and research staff consulted over the last few years by one of the authors.
9. Much of the critique of limited democratic and participatory practice within community-based or farmers' organizations is rarely published but is often voiced among academics, NGOs, donors, and policymakers; for cautiously critical comments, see also Tendler 1993 and contributions to Bebbington and Thiele 1993.
10. Putnam (1995), quoted in Ritchey-Vance, Chapter 3 of this book. Also see MacGillivray, Lingayah, and Zadek 1997 for an extensive discussion about the links between social capital and the social economy.
11. From discussions with the author.
12. See Tendler (1993) for Brazil and Biggs and Smith (1995) for a discussion of social dynamics within programs and institutions aiming at participatory agricultural research and development practices in different contexts. Blauert and Quintanar (1997) present a first analysis of their evaluation work carried out on project internal processes and qualitative aspects of a farmer-to-farmer project.
13. Many serial publications and sources of gray literature make accessible examples

of work in the arena of traditional agricultural knowledge and farming systems and of participatory technology development and sustainable agricultural production or general resource use; most of these publications also have the advantage of providing cases from other regions of the world, to set the Latin American context into a global context. See, for example, *ILEIA Newsletter, Forest, Trees and People* (FAO), IT Publications (London), working chapters by the Overseas Development Institute (ODI) in London, edited books such as Scoones and Thompson 1994, and the Gatekeeper Series of the International Institute of Environment and Development (IIED), Sustainable Agriculture Programme, London.

References

Anderson, Simon. 1996. Research Centers and Participatory Research: Issues and Implications (Mexico). Revised paper to the Study Group on Mediating Sustainability, Institute of Latin American Studies, London.

Barratt Brown, Michael. 1993. *Fair Trade*. London: Zed Books.

Bebbington, Anthony, and Graham Thiele. 1993. *NGOs and the State in Latin America: Rethinking Roles in Sustainable Agricultural Development*. London: Routledge.

Biggs, Stephen, and Grant Smith. 1995. Contending Coalitions in Agricultural Research and Development: Challenges for Planning and Management. Chapter for the conference Evaluation for a New Century: A Global Perspective, Canadian Evaluation Association and American Evaluation Association, Vancouver, Canada, 1–5 November.

Blauert, Jutta, and Eduardo Quintanar. 1997. *Seeking Local Indicators: Participatory Stake-holder Evaluation of Farmer-to-Farmer Projects in Southern Mexico*. PLA Notes no. 28. London: IIED.

Calva, José Luis. 1993. Principios Fundamentales de un Modelo de Desarrollo Agropecuario Adecuado para México. In *Alternativas para el Campo Mexicano*, vol. 2, edited by J. L. Calva. Mexico City: Fundación Friedrich Ebert/UNAM.

Does, Marcel vander, and Alberto Arce. 1995. The Use of Narrative in Project Evaluation: A Case from Ecuador. Revised paper presented to the Study Group on Mediating Sustainability, Wageningen, Netherlands, November.

Doyal, Len, and Ian Gough. 1991. *A Theory of Human Needs*. London: Macmillan.

Edwards, Michael, and David Hulme, eds. 1995. *Non-Governmental Organizations —Performance and Accountability*. London: Earthscan.

Escalante, Roberto. 1996. Participation and Economics: Illustrations from Mexico's Agricultural Sector. Revised paper presented to the Study Group on Mediating Sustainability, Institute of Latin American Studies, London.

FAO. 1994. *Participación Campesina para un Agricultura Sostenible en Países de América Latina*, Series "Participación Popular" no. 7. Rome: FAO.

Fox, J., and J. Aranda. 1996. *Decentralization and Rural Development in Mexico: Community Participation in Oaxaca's Municipal Funds Program*. Monograph Series 42. San Diego: Center for U.S.-Mexico Studies, University of California.

Gómez C., R. Manuel Angel, R. Schwentesius, M. Muñoz Rodríguez, et al. 1993. ¿PROCAMPO ó ANTICAMPO? CIESTAM. Universidad Autónoma de Chapingo, Reporte de Investigación no. 20, Mexico City.

Gubbels, Peter. 1994. Populist Pipe-dream or Practical Paradigm? Farmer-Driven

Research and the Project Agro-forester in Burkina Faso. In *Beyond Farmer First*, edited by Ian Scoones and John Thompson. London: IT Publications.

Holt-Giménez, Eric. 1995. The Campesino-a-Campesino Movement: Farmer-Led Agricultural Extension. IIRR/ODI Workshop on Farmer-Led Approaches to Agricultural Extension, Philippines, 17–22 July.

Linck, Thierry, ed. 1993. *Agriculturas y Campesinados de América Latina. Mutaciones y Recomposiciones.* Mexico City: Fondo de Cultura Económica/ORSTOM.

MacGillivray, Alex, Sanjit Lingayah, and Simon Zadek. 1997. *Social Policy as a Productive Factor.* London: New Economics Foundation.

MacGillivray, Alex, and Simon Zadek. 1996. Medir la Sostenibilidad: Revisíon Sobre el Arte de Hacer que Funcionen los Indicadores. *Investigación Económica* 56 (218): 139–76.

Max-Neef, Manfred. 1991. *Human Scale Development: Conceptions, Application and Further Reflections.* With contributions from Antonio Elizalde and Martin Hopenhayn. New York and London: Apex Press.

Niekerk, N. van. 1994. *Desarrollo Rural en los Andes. Un estudio sobre los programas de desarrollo de Organizaciones No Gubernamentales.* Leiden Development Studies no. 13. Faculty of Social Sciences, Leiden University, Leiden, Netherlands.

Péres, José Antonio. 1996. Reforms, Actors and Popular Participation in Contemporary Bolivia. Paper presented to the Study Group on Mediating Sustainability, Wageningen. Revised version.

Putnam, Robert. 1995. Social Capital, *People Centred Development Forum Newsletter*, 6 March, column 76.

Ruben, Ruerd, and Niko Heerink. 1996. Economic Approaches for the Evaluation of Low External Input Agriculture. Draft paper, Study Group on Mediating Sustainability, London. (For a shorter version, see Economic Evaluation of LEISA Farming. *ILEIA Newsletter* 11 [2]: 18–20.

Scoones, Ian, and John Thompson, eds. 1994. *Beyond Farmer First: Rural People's Knowledge, Agricultural Research and Extension Practice.* London: IT Publications.

Sogge, David, ed. 1996. *Compassion or Calculation: The Business of Private Foreign Aid.* London: Pluto Press/Trans-national Institute.

Tendler, Judith. 1993. Tales of Dissemination in Small-Farm Agriculture: Lessons for Institution Builders. *World Development* 21 (10): 1567–82.

Thrupp, Lori Ann, ed. 1996. *New Partnerships for Sustainable Agriculture.* Washington, D.C.: World Resources Institute.

Toranzo, Carlos. 1995. Los Retos para Las ONG. *IDC* 2:6.

UNDPI. 1993. Agenda 21: The United Nations Program of Action from Rio. New York: UN.

World Bank. 1994. *The World Bank and Participation.* Washington, D.C.: IBRD Operations Policy Department.

Zadek, Simon. 1995. *Value-Based Organization for Effectiveness.* London: New Economics Foundation.

Zadek, Simon, and Alex MacGillivray. 1996. Measuring Sustainability: From Local Diversity to Macro Focus. Revised paper presented to the Study Group on Mediating Sustainability, London.

Zadek, S., and P. Tiffen. 1996. Fair Trade: Business or Campaign. *Development* 1996 (3): 48–53.

Abbreviations

ACP	African, Caribbean, and Pacific
ACPEC	African, Caribbean, and Pacific Economic Conference
ADC	Alianza Democrática Campesina (El Salvador)
AECO	Asociación Ecológica Costaricense
AIDS	acquired immunodeficiency syndrome
AMA	Acuerdo Mexicano-Alemán
ANED	Asociación Nacional Ecuménica del Desarrollo (Bolivia)
APEMEP	Asociación de Pequeños y Medianos Productores de Panamá
AS-PTA	Assessoria e Servicos a Projetos en Agricultura Alternativa (Brazil)
ASEPROLA	Asociación de Servicios de Promoción Laboral
ASOCODE	Asociación Centroamericana de Organizaciones Campesinas para la Cooperación y el Desarrollo
ASOTRAMA	Asociación de Trabajadores y Medio Ambiente
ATC	Asociación de Trabajadores del Campo
ATO	alternative trading organization
BFAC	Belize Federation of Agricultural Cooperatives
BGA	Banana Growers' Association
BL	Banana Link
CADESCA	Comité de Apoyo al Desarrollo Económico y Social de Centroamérica (Guatemala)
CAP	Compromiso Agropecuario de Panamá (1992)
CARE-UNA	Proyecto CARE y Universidad Nacional Agrícola
CARICOM	Caribbean Common Market
CASA (DK)	Danish NGO program
CBO	community-based organization
CCC-B	Confederation of Cooperatives and Credit Unions of Belize
CCC-CA	Confederación de Cooperativas del Caribe y Centroamérica
CCJYD	Consejo Campesino Justicia y Desarrollo (Costa Rica)
CCMSS	Consejo Civil Mexicano para le Silvicultura Sostenible
CCOD	Concertación Centroamericana de Organismos de Desarrollo
CECADE	Centro de Capacitación para el Desarrollo
CEE	Comunidad Económica Europea (EEC)
CENCAP	Centro de Capacitación Cooperativista (El Salvador)
CEPA	State Planning Institute (Brazil)
CEPAL	Comisión Económica para América Latina
CEPB	Centro de Empresa Privada Boliviana
CEPCO	Coordinadora Estatal de Productores de Café de Oaxaca (Mexico)

CEPLAES	Centro de Planificación y Estudios Sociales (Ecuador)
CERES	Dutch Social and Economic Science Research Council
CGIAR	Consultative Group on International Agricultural Research
CIAT	Centro de Investigaciones Agrícolas del Trópico (Bolivia)
CICAFOC	Coordinadora Indígena Campesina de Agroforestería Comunitaria Centroamericana
CIFCA	Copenhagen Initiative for Central America (1991)
CIMMYT	International Center for Maize and Wheat Improvement (Mexico)
CIP	International Potato Center (Peru)
CIPCA	Centro de Investigación y Promoción Campesina (Bolivia)
CIPRES	Centro de Investigación, Promoción y Desarrollo Rural y Social (Nicaragua)
CNA	Coordinadora Nacional Agraria (Costa Rica)
CNC	Confederación Nacional Campesina (Mexico)
CNC	Consejo Nacional Campesino (Honduras)
CNOC	Coordinadora Nacional de Organizaciones Cafetaleras (Mexico)
CNP	Consejo Nacional de Producción (Costa Rica)
CNTC	Central Nacional de Trabajadores del Campo (Honduras)
COACES	Confederación de Asociaciones Cooperativas de El Salvador
COCENTRA	Coordinadora Centroamericana de Trabajadores
COCOCH	Concejo Coordinador de Organizaciones Campesinas de Honduras
COMUNIDEC	community development organization (Ecuador)
CONAC	Confederación Nacional de Asentamientos Campesinos (Panama)
CONAMPRO	Coordinadora Nacional de Pequeños y Medianos Productores (Guatemala)
CONFRAS	Confederación de Federaciones de Cooperatives de la Reforma Agraria Salvadoreña
CONIC	Coordinadora Nacional Indígena Campesina (Guatemala)
COOPEAGRI	Cooperativa Agrícola Industrial y de Servicios Múltiples (Costa Rica)
COSECHA	Asociación de Consejeros para una Agricultura Sostenible, Ecológica y Humana (Honduras)
CRASX	Consejo Regional Agrosilvo-pastoril y de Servicios X'pujil (Mexico)
CRIES	Centro Regional de Investigaciones Económicas y Sociales (Nicaragua)
CSD	Commission for Sustainable Development
CTA-ZM	Centro de Tecnologías Alternativas da Zona da Mata (Brazil)
CTM	Central de Trabajadores Mexicanos
CUC	Comité de Unidad Campesina (Guatemala)
DEI	Departamento Ecuménico de Investigaciones (Costa Rica)
DESCO	Centro de Estudios y Promoción del Desarrollo (Peru)
DG	directorate general

DGDF	Dirección General de Desarrollo Forestal (Mexico)
DPH	Diálogo para el Progeso Humano
DRPA	Diagnóstico Rápido Participativo da Agroecosistema
DRU	Grupo de Desarrollo Rural (Bolivia)
DTF	Dirección Técnica Forestal (Mexico)
EDPYMEs	Entidades de Desarrollo para la Pequeña y Mediana Empresa
EDUCA	Editorial Universitaria Centroamericana
EMATER	state-level extension service (Nordeste, Brazil)
ENABAS	Empresa Nicaragüense Productos Básicos (Nicaragua)
ERA	Estudios Rurales y Asesoría (Mexico)
EU	European Union
EUROBAN	European Banana Action Network
FAC	Foro Agropecuario Campesino Centroamericano
FACA-UNA	Facultad Agrícola Universidad Nacional Agrícola (Nicaragua)
FAO	Food and Agriculture Organization
FEDIPRICAP	Federación de Entidades Privadas de Centroamérica y Panamá
FENACOOP	Federación Nacional de Cooperativas (Nicaragua)
FESACORASAL	Federación de Cooperativas de la Reforma Agraria de la Región Occidental (El Salvador)
FFPs	Fondos Financieros Privados
FIFSIDA	Foundation for Initiatives Against AIDS
FIPA	Federación Internacional de Productores Agropecuarios (France)
FL	Farmers' Link
FLO	Fairtrade Labelling Organisation
FMLN	Frente Farabundo Martí para la Liberación Nacional (El Salvador)
FOB	free on board
FONDAD	Foro sobre la Deuda y el Desarrollo
FONDECO	Fondo del Desarrollo Comunal
FSLN	Frente Sandinista de Liberación Nacional (Nicaragua)
FTMO	fair-trade marketing organization
FUNDE	Fundación Nacional para el Desarrollo (El Salvador)
FUNDESCA	Fundación para el Desarrollo Económico y Social de Centroamérica (Panama)
GATT	General Agreement on Tariffs and Trade
GDP	gross domestic product
GEA	Grupo de Estudios Ambientales (Mexico)
GIA	Grupo de Investigaciones Agrícolas (Chile)
GLASOD	The Global Assessment of Human Induced Soil Degradation
GSO	grassroots support organization
GTZ	Gesellschaft für Technische Zusammenarbeit
HIV	human immunodeficiency virus
IAF	Inter-American Foundation
IAT	International Agribusiness Trading Corporation

IBRD	International Bank for Reconstruction and Development
ICA	International Coffee Agreement
ICIC	Iniciativa Civil para le Integración Centroamericana
ICO	International Coffee Organization
ICS	in-country support
ICTUR	International Centre for Trade Union Rights
IDB	Inter-American Development Bank
idc	Instituciones, Desarrollo, Cooperación
IDS	Institute of Development Studies, University of Sussex
IFAD	International Fund for Agricultural Development
IFAT	International Federation for Alternative Trade
IFPRI	International Food Policy Research Institute
IICA	Inter-American Institute for Agricultural Cooperation (Costa Rica)
IIED	International Institute of Environment and Development
ILO	International Labor Organization
IMF	International Monetary Fund
INDAP	Instituto de Desarrollo Agropecuario (Chile)
INIBAP	International Network for the Improvement of Bananas and Plantains
INMECAFE	Mexican coffee parastatal
IPC	Interdisziplinäre Project Consult
IPE	Información Política y Económica
IRA	Instituto Regulador de Abastecimientos (El Salvador)
ISEC	Instituto de Sociología y Estudios Campesinos
ISNAR	International Service for National Agricultural Research
ISO	International Standards Organization
LAC	Latin American and Caribbean
LEEC	London Environmental Economics Centre
M&E	monitoring and evaluation
MAI	Multilateral Agreement on Investment
MCCH	Comercializando como Hermanos
MHF	Max Havelaar Foundation
MOCAF	Red Mexicana de Organizaciones Campesinas Forestales
NAFTA	North American Free Trade Agreement
NARS	National Agricultural Research System
NEF	New Economics Foundation
NGO	nongovernmental organization
NOGUB	nongovernmental organizations program of Swiss aid in Bolivia
NSM	new social movement
OCD	community development organization
OCIA	Organic Crop Improvement Association
ODI	Overseas Development Institute
OECD	Organization for Economic Cooperation and Development
OEPF-ZM	Organización de Ejidos Productores Forestals Zona Maya (Mexico)

OFCOR	on-farm client-oriented research
OPAM	Organisation Patriotique des Agricultuers Martiniquais
OPEC	Organization of Petroleum Exporting Countries
PAC	Plan de Acción para la Agricultura Centroamericana
PAR	participatory action research
PARLACEN	Parlamento Centroamericano
PASONAC	Central American Program of Sustainable Agriculture
PASOS	Programa Practicas de Desarrollo Rural (Grupo de Estudios Ambientales) (Mexico)
PDVSA	national petroleum company of Venezuela
PFSA	Programma de Formación en Seguridad Alimentaria
PIM	participatory impact monitoring
PIP	Plano Individual da Propriedade
PME	participatory monitoring and evaluation
PNMH	Program for River Microcatchments (Brazil)
PNUD	Programa de las Naciones Unidas para el Desarrollo
PPF	Plan Piloto Forestal (Mexico)
PRA	Participatory Rural Appraisal
PRI	Partido Revolucionario Institucional
PROCAMPO	Program of Support to the Countryside (Mexico)
PRODER	Program of Regional Development (Mexico)
PROSOLO	Soil Conservation, Management, and Pollution Control Incentive Program
PSA	Programa de Seguridad Alimentaria
RIAD	Red Interamericana de Agricultura y Democracia
RONGEAD	Reseau des ONGs Europeennes pour l'Agriculture et le Developpement
RRA	Rapid Rural Appraisal
SAED	Servicios Alternativos para la Educación y el Desarrollo
SAP	structural adjustment program
SARD	sustainable agriculture and rural development
SARH-INIF	Agricultural Ministry and National Foresty Research Institute (Mexico)
SCFS	Small Farmers Cooperative Society
SEDEPAC	Servicio para el Desarrollo y Paz (Mexico)
SEDESO	Social Development Ministry (Mexico)
SELA	Sistema Económico Latinoamericano
SEMARNAP	Environment Ministry (Mexico)
SERRV	U.S. fair-trade nonprofit organization
SICA	Sistema de la Integración Centroamericana
SIMAS	Sistema de Información Mesoamericano sobre Agricultura Sostenible
SINDICARNE	association of poultry- and pig-producing industries
SINDIFUMO	association of tobacco-producing industries
SOAS	School of Oriental and African Studies
SOCRA	Sociedad de Cooperativas Cafetaleras de la Reforma Agraria (El Salvador)

STR	Sindicato dos Trabalhadores Rurais
SWC	soil and water conservation
TAC	Technical Advisory Committee
TANICA	Tanzanian coffee parastatal
TNI	Trans-National Institute
TWIN	Third World Information Network
UACh	Universidad Autónoma de Chapingo (Mexico)
UAM	Universidad Autónoma de Mexico
UCADEGUA	Unión Campesina de Guatuso (Costa Rica)
UCIRI	Unión de Campesinos Indigenas de la Región del Istmo (Mexico)
UCS	Unión Comunal Salvadoreña
UN	United Nations
UNAG	Unión Nacional de Agricultores y Ganaderos (Nicaragua)
UNAM	Universidad Nacional Autónoma de México
UNC	Unión Nacional de Campesinos (Honduras)
UNCAFESUR	Unión de Cafeticultores del Sur (Mexico)
UNDP	United Nations Development Program
UNDPI	United Nations Department of Public Information
UNED-UK	U.K. program for United Nations Environment and Development activities
UNEP	United Nations Evnironmental Program
UNESCO	United Nations Scientific, Educational, and Cultural Organization
UOCACI	Unión de Organizaciones Campesinas de Cicalpa (Ecuador)
UPANACIONAL	Unión Nacional de Pequeños y Medianos Productores Agricolas (Costa Rica)
UPEB	Unión de Paises Exportadores del Banano
UPMP	Unidades Productoras de Materia Prima Forestal (Mexico)
UPROCAFE	Unión de Pequeños y Medianos Productores de Café de Centro América, México y del Caribe
UROCAL	Unión Regional de Organizaciones Campesinas del Litoral
USAID	U.S. Agency for International Development
USDA	U.S. Department of Agriculture
USTR	United States trade representative
UZACHI	Unión Zapoteca y Chinanteca del Istmo (Mexico)
WHO	World Health Organization
WIBDECO	Windward Islands Banana Development and Exporting Company
WINFA	Windward Islands Farmers' Association
WN	World Neighbors
WRI	World Resources Institute
WTO	World Trade Organization
WWF	World Wildlife Fund

Index